GLOBAL WARMING – THE RESEARCH CHALLENGES

GLOBAL WARMING – THE RESEARCH CHALLENGES

A Report of Japan's Global Warming Initiative

Edited by

ATSUNOBU ICHIKAWA
*Council for Science and Technology Policy,
The Government of Japan, Tokyo, Japan
and
Professor Emeritus,
Tokyo Institute of Technology,
Tokyo, Japan*

 Springer

Library of Congress Cataloging-in-Publication Data

ISBN 1-4020-2940-3 (HB)
ISBN 1-4020-2941-1 (e-book)

Published by Springer,
P.O. Box 17, 3300 AA Dordrecht, The Netherlands.

Sold and distributed in North, Central and South America
by Springer,
101 Philip Drive, Norwell, MA 02061, U.S.A.

In all other countries, sold and distributed
by Springer,
P.O. Box 322, 3300 AH Dordrecht, The Netherlands.

First published 2003, under the title "Chikyu Ondanka Kenkyu no Saizensen" (The State of the Art of Global Warming Research), ISBN 4-17-264130-X, by Printing Bureau, Ministry of Finance, Japan

Printed on acid-free paper

springeronline.com

All Rights Reserved
© 2004 Springer
No part of this work may be reproduced, stored in a retrieval system, or transmitted
in any form or by any means, electronic, mechanical, photocopying, microfilming, recording
or otherwise, without written permission from the Publisher, with the exception
of any material supplied specifically for the purpose of being entered
and executed on a computer system, for exclusive use by the purchaser of the work.

Printed in the Netherlands.

Table of Contents

FOREWORD vii

PREFACE TO THE JAPANESE EDITION xi

PART 1:
ADDRESSING THE PROBLEM OF GLOBAL WARMING

1. A New Era of Global Environment Issues 3
2. Answering Questions About Global Warming 4
3. Top Priority Research and Development in the Field of Environmental Studies 5
4. Global Warming Research Initiative 9

PART 2:
TO WHAT EXTENT HAVE RESEARCHES ELUCIDATED GLOBAL WARMING?

CHAPTER 1
IS GLOBAL WARMING REALLY OCCURRING?
– WHAT GLOBAL MONITORING CAN TELL US 19

1.1 Introduction 19
1.2 How Has the Earth's Climate Changed? 20
1.3 Atmospheric Concentration of Greenhouse Gases and Aerosols: Influence on Past, Present, and Future Climate 26
1.4 Terrestrial and Oceanic Sources and Sinks for Major Greenhouse Gases 33

CHAPTER 2
CLIMATE MODELING AND THE PROJECTION OF
GLOBAL WARMING 55

2.1 History of the Projection of Global Warming 55
2.2 How is Global Warming Actually Projected? 58
2.3 Worldwide Research Efforts on Projecting Global Warming 63
2.4 National Efforts in Projection Research on Global Warming 66
Notes 83

CHAPTER 3
IMPACTS AND RISKS OF GLOBAL WARMING 85

3.1 Global Impacts 85
3.2 Impacts on Japan 100

CHAPTER 4
ASSESSMENT OF GLOBAL WARMING RESPONSE
POLICIES 115

4.1 Introduction 115
4.2 Formulating Hundred-Year Scenarios 117
4.3 Estimated Costs of the Kyoto Protocol 121
4.4 Technological Innovation Making Greater Progress Than
    Expected 124
4.5 Policy Design Growing in Sophistication 127
4.6 New Insights for the Rules of Consensus 129
4.7 The Difficulty of Judging the Balance of Losses and Gains from
    Mitigation Measures 131
4.8 Response Policy Studies Entering a New Phase 134
Notes 137

REFERENCES 141

ABBREVIATIONS 151

AUTHORS 153

INDEX 157

FOREWORD

This monograph is the English version of *"Chikyu Ondanka Kenkyu no Saizensen* (The State of the Art of the Global Warming Research)" published as a report of Global Warming Research Initiative established by The Council for Science and Technology Policy (CSTP), the Cabinet Office, Japanese Government.

CSTP promulgated the "Promotional Strategy in Prioritized Fields" in September 2001 in accordance with the "Science and Technology Basic Plan", now in its second term, which started fiscal year 2001 and will continue through 2005. CSTP organized "Research Initiatives" in the field of environmental studies to identify and integrate research activities that had been separately conducted under the auspices and the directions of many ministries and agencies in the Japanese government. CSTP has established five research initiatives in the field, i.e., Global Warming, Waste-Free and Resource Recycling Technologies, Eco-Harmonious River Basin and Urban Area Regeneration, Chemical Substance Risk Management, and Global Water Cycle.

The Global Warming Research Initiative (GWRI) consists of two study areas, the Climate Change Research Area (CCRA) and Mitigation Technology Development Area. The objective of CCRA is clear; it is being undertaken to acquire scientific knowledge regarding the effect of anthropogenic emissions of greenhouse gases on climate change. The ultimate goal of CCRA is to establish a set of comprehensive and integrated models of greenhouse gas emissions with mitigation measures as the input and climate change as the output. We can then determine the amounts of greenhouse gas emissions allowable to stabilize climate change. Clarification and integration of research activities are particularly significant in this area since formation of a unified model is crucial to predict the future climate, to assess the impact of climate change, and to adopt appropriate political, economic, and technological measures. GWRI/CCRA thus includes four research programs: Monitoring and Process Study, Projection Modeling and Climate Change Study, Impact and Risk Assessment, and Response Policies.

The promotion strategy of CSTP places specific emphasis on dissemination of research outcomes of environmental studies to people in communities other than science, to not only ensure further support from the people in these communities but also to provide accurate information about coping with future changes in the global environment. The Council Member in charge of the environmental area has been encouraging the five research initiatives to publish research reports with the above intention.

GWRI/CCRA decided at its first project leader meeting, held on April 10, 2002, to report the outcome of the research initiative to the general public, policy makers, and the research community. The Steering Committee of CCRA responded to this decision by discussing the objective, constitution, and contents of the research report. It was decided during this discussion that the report would act as the centripetal force for the researchers participating in CCRA. The following three objectives of the report were therefore defined:

(1) To describe climate change research in a comprehensive form to the general public, which is concerned with environmental issues,
(2) To transmit the research results to the legislature and government officials in a comprehensive form so as to become a sound foundation for policy decisions, and
(3) To provide an extensive view of researchers engaged in studying climate change areas other than their specialties.

This report describes global and domestic research outcomes, which will become common knowledge for all researchers engaged in GWR. The report was published in 2002 under the supervision of the CSTP member in charge of the environmental area and the Director General for Science and Technology in the Cabinet Office, since the objectives are consistent with those of CSTP.

This English book is based on the original *"Chikyu Ondanka Kenkyu no Saizensen"* with new information added to reflect the latest progress of research. The contents of the book are as follows.

Part 1 details the purpose, system, and organization of GWRI/CCRA, which is organized as an inter-ministerial research activity.

Part 2 describes existing research outcomes for monitoring, observation, and process analyses (Chapter 1), the climate model and simulations (Chapter 2), assessment of impacts of warming (Chapter 3), and mitigation and countermeasure policies (Chapter 4). The underlying intention of all chapters is to examine to what extent the global warming mechanism has been clarified.

This book has been made possible through great efforts by many people involved in GWRI/CCRA.

Prof. Shiro Ishii, former CSTP member in charge of environmental studies, supervised from the CSTP viewpoint. The CSTP Secretariat in charge of the environment and energy, headed by Dr. Makoto Watanabe, former Director for Environment and Energy, Dr. Yasuhiro Sasano, and

Director for Environment and Energy contributed substantially to the creation of this report.

Professor and Director Isao Koike, Ocean Research Institute, the University of Tokyo; Professor Nobuo Mimura, Ibaraki University; Dr. Akira Noda, Meteorological Research Institute; Dr. Hiroki Kondo, Frontier Research System for Global Change and Dr. Hideo Harasawa, National Institute for Environmental Studies, contributed greatly by serving as members of the Editorial Committee. Among them, Professor Nobuo Mimura served as arbitrator, a de facto chair of the committee.

The authors of this book, whose names are provided in the appendices, were challenged by the difficult task of composing manuscripts at a scientific level that could be easily understood by all readers. I would like to express my heartfelt appreciation to all the above-mentioned people.

I sincerely hope that this book is helpful for those among the general public, policy-makers, and researchers who are concerned with the global environment. I welcome constructive comments and suggestions regarding GWRI/CCRA.

February 2004
Atsunobu Ichikawa, Chair
Climate Change Research Area,
Global Warming Research Initiative
Council for Science and Technology Policy

PREFACE TO THE JAPANESE EDITION

The Council for Science and Technology Policy (CSTP) is an organization that was inaugurated in January 2001 as one of the important policy councils set up within the Cabinet Office. It is envisioned as functioning as a control tower for policy promotion in the area of science and technology. In view of this mandate, CSTP is formulating "Promotional Strategies in Prioritized Fields" stipulated by the Science and Technology Basic Plan, which is now in its second phase (2001 through 2005). Five research topics as priority issues in the environmental field have been positioned within the strategies. These five spheres of research are (1) global warming, (2) waste-free and resource recycling technologies, (3) eco-harmonious river basin and urban area regeneration, (4) chemical substance risk management and (5) global water cycle. Currently, various governmental ministries and agencies are conducting separate research in the environmental field. Those individual research projects are to be codified and restructured to obtain a coherent view and are to be promoted according to scenario-driven initiatives that will set a course by which the government as a whole will work toward and achieve the same policy objectives.

Talk about eliminating ministerial sectionalism and promoting collaboration among government offices invariably surfaces in connection with virtually all matters. Furthermore, the promotion of initiative structure in particular adopted for research related to priority issues in the environmental field is none other than the nature of environmental research itself.

It goes without saying that this genre of research is performed to solve environmental problems. In order to do that, though, the collective wisdom of humankind must be mobilized. While such wisdom of course encompasses different scientific and technical domains, it also incorporates the humanities and social sciences. Accumulating bottom-up research is not the answer. Instead, it is imperative to construct and put into operation a structure that decisively designates the course to be taken and processes for solving problems (a scenario), that draws on various elements of science and technology in accordance with that scenario, and that serves as a driving force for progress.

In actuality, CSTP started to set this initiative structure in motion during the current fiscal year (FY 2002). One thing that was keenly felt in the process was the need to check the validity of the path for addressing problems. The status of developments in research under an initiative must be kept firmly under control at each juncture and compared with the scenario.

There are multiple ways of doing this. It goes without saying, for example, that liaison committees, study groups, workshops, and symposiums are effective and productive options. However, there is probably another way that is becoming more important than anything else: written descriptions of the status of research progress that are presented in the form of structured reports. For it is through such reports that initiative activities and results are made available to the public in an objective format. These reports are directly related to accountability to taxpayers, something that is much talked about. At the same time, through these reports the parties who are involved also personally realize the depth of problems and the arduous nature of the path to solving them. Moreover, they communicate that to the public.

As everyone knows by now, environmental problems are issues of a dimension that involves the way of life of humankind as a whole. Within that context, the scientific and technological community must maintain communication by conveying messages about the environmental research entrusted to it by society, which then once again generates research assignments. It is this point, communication, that is the intention behind the publication of this report.

With this year being the newly launched initiative structure's first year, we are not yet at the stage at which new research results can be presented. Consequently, in connection with the fact that this is the year of the Johannesburg Summit, we decided to sort out and explain developments along the international front lines of research into the phenomenon of global warming. This report is not simply a description of the current status of research in this field. Instead, it confirms, so to speak, the situation at the point when the initiative for global warming research was launched and provides a reference line for evaluating the performance of activities in the future, with results constantly being compared with this benchmark.

Naturally, given the aforementioned intention behind this report, we expect to continue publishing subsequent editions in the years to come. The themes they will cover and the format each will take will be determined according to the status of progress with regard to initiatives related to other problems. It is our hope that, with each passing year, this series of reports will be increasingly well received by the public and will be able to contribute to the advancement of society's efforts to address environmental problems.

December 2002
Shiro Ishii, Member, Council for Science and Technology Policy.
Kenji Okuma, Director General for Science and Technology Policy, Cabinet Office, Government of Japan.

PART 1

ADDRESSING THE PROBLEM OF GLOBAL WARMING

1. A NEW ERA OF GLOBAL ENVIRONMENT ISSUES

The year 2002 will be remembered as a turning point in addressing global environment problems in Japan; a new era of resolving the issues had just begun.

Japan ratified the Kyoto Protocol, the United Nations Framework Convention on Climate Change (UNFCCC), on June 4, 2002, as the 74th party to the protocol. As a result, Japan is legally obliged to reduce greenhouse gas emissions by 6%, converted into CO_2 equivalent, of the 1990 level.

The year 2002 is also notable as the year when the World Summit on Sustainable Development was held in Johannesburg, South Africa (Johannesburg Summit 2002). Leaders from 191 countries and regions throughout the world met at the Johannesburg Summit to confront the challenge of globally sustainable development in harmony with the environment. More than 20,000 attendants, including leaders of non-governmental organizations (NGOs) and representatives from industries and major international organizations, also convened at the conference. Stable food supplies, potable quality water, housing, sanitary conditions, energy, medical service, and economic stability are becoming increasingly important as the world population increases. Various problems were discussed at the Johannesburg Summit, including improving the quality of life for all and conserving our natural resources. The Johannesburg Declaration was then agreed upon by participating countries.

Participating countries adopted a global plan of action for sustainable development, Agenda 21, at the United Nations Conference on Environment and Development (UNCED) held in Rio de Janeiro, Brazil, in 1992. The measures set forth under Agenda 21 were reviewed at the Johannesburg Summit held ten years after the Rio Summit, and a practical program for more effective implementation, the "Execution Plan," was created. A Commitment Document describing the plan of action proposed by individual countries, international organizations, and NGOs was also published.

The year 2002 also holds special significance as an epoch-making year in the environmental research field in Japan. The "Environmental Initiative," a Japanese research frame for promoting inter-disciplinary research among governmental ministries and agencies in the environmental field and one of the four strategic fields defined by the Council for Science and Technology Policy (CSTP), Cabinet Office (see Section 3), was initiated. The impenetrable walls around governmental ministries and agencies have traditionally hindered the development of science and technology research. The framework of the Environmental Initiative was established to solve this problem in environmental research, which is categorized as a problem-solving research field.

2. ANSWERING QUESTIONS ABOUT GLOBAL WARMING

The topics of global warming often dominate TV news programs and fill the pages of newspapers and magazines, and many questions have emerged. Is the earth really warming? Are CO_2 emissions the true cause of global warming? Will global warming worsen and cause irrecoverable damage? And if so, how can such problems be avoided? What preventive measures can be taken?

These questions are natural and common to all people. How will scientists answer these questions? The questions regarding global warming can be summarized as follows.

- What are the facts regarding climate and the concentration of greenhouse gases?
- What degree of climate change is acceptable?
- How will anthropogenic greenhouse gas emissions change?
- Where and how will climate changes appear?
- How can technology reduce greenhouse gas emissions?
- Can CO_2 be collected and fixed?
- When should action be taken and of what type?
- To what extent should greenhouse gas emissions be suppressed?

The Global Warming Research Initiative is a research framework that has been set up to answer these questions. This book is aimed at clarifying the significance and objectives of the initiative by reviewing the present status of international and domestic research on climatic change. It also reveals what science can and cannot answer among the questions listed above.

3. TOP PRIORITY RESEARCH AND DEVELOPMENT IN THE FIELD OF ENVIRONMENTAL STUDIES

Council for Science and Technology Policy Formed by Administrative Reform

The Japanese governmental administrative system was reformed in January 2001. The Prime Minister's Office and 22 ministries and agencies were transformed into a new scheme consisting of the Cabinet Office and 12 ministries. A new Council for Science and Technology Policy (CSTP) was established within the Cabinet Office as one of the government top councils with the mission of important policy-making under the Incorporation Act of the Cabinet Law. Its objectives are basic, comprehensive policy-making for science and technology and general coordination. CSTP takes the initiative from among all ministries, with an overall view from one level higher than those of other ministries, thus serving as a source of wisdom to support the Prime Minister and the Cabinet.

Role of the Council for Science and Technology Policy

CSTP is expected to eliminate divided administrative functions and act as a leader with strong vision and responsiveness to promote broad strategic scientific and technological policies in Japan under the initiative of the chairman, the Prime Minister. CSTP established basic strategies for science and technology and policies for budget and human resource allocations for scientific and technological research to meet national challenges, such as revitalizing the economy, dealing with the problem of the aging of society, and various problems, including global warming. Key national research and development (R&D) projects are identified and evaluated by CSTP.

Photo 1. The Council for Science and Technology Policy chaired by Prime Minister Koizumi (November 11, 2002). The Council consists of the Chair and 14 or less members. Upper: Member Ministers, lower: Invited Experts.

Promotional Strategies in Prioritized Fields

The Second Basic Plan for Science and Technology (2001-2005) enacted in 2001 describes a policy of prioritization in R&D. CSTP arranged an "Expert Panel on Promotion Strategies of Prioritized Areas" to implement this plan. The panel focuses on four fields: life science, information technologies, the environment, and nanotechnology. Also prioritized were the four fields of energy, manufacturing technology, social infrastructure, and space science and marine frontiers, which are necessary for building a strong foundation for a viable country. Objectives were clarified, and policies for five-year R&D plans were set through Initiative investigations by project teams for these fields.

The discussion results were compiled as "Promotional Strategies in Prioritized Fields" and adopted by the CSTP Plenary Session in September 2001.

Strategies for Environmental Research –Research Initiatives–

Leading researchers in the field and experts from the private sectors joined with responsible CSTP members and discussed themes to formulate firm strategies for environmental research. The results are summarized as follows.

Existing and potential environmental problems are extremely complex and diverse. Therefore, research on individual phenomena conducted separately by ministries and institutes cannot solve the problems effectively. It is important to establish a comprehensive framework for promoting research across existing academic boundaries. Environmental R&D policies must foster coordination and cooperation among ministries and between industry and academia both domestically and internationally. While some research programs are carried out with the cooperation of several ministries, individual ministries and institutions often conduct research independently, thus hindering the integration of individual research results.

Elimination of the administrative sectionalism requires a review of existing strategies and programs for environmental research and technological development conducted by individual ministries and the establishment of initiatives to promote scenario-driven R&D. The political objectives are defined from the general standpoint of the government, and the roadmap toward them should be clear. In addition, appropriate allocation of roles and collaboration should be pursued among industry, government, and academia.

An initiative framework provides the following advantages.
- National administrative efforts and the methods for solving problems will be easily understood not only by researchers, but by the government and the public.
- Duplicate R&D will be eliminated and systematic cooperation will be established. Well-organized research results will be used to set environmental policies.
- Japan will act as a leader in international efforts to resolve problems.

Five research initiatives were established in the following environmental field:
(a) Global Warming
This initiative achieves the final UNFCCC goal by providing

scientific knowledge and technology for successful policy making through international cooperation.

(b) Waste-Free and Resource Recycling Technologies

This initiative develops technologies to reduce waste production and minimize environmental risks to realize the goals of the Basic Law for Establishing a Recycling-based Society and to establish an appropriate recycling system in collaboration with overseas countries.

(c) Eco-Harmonious River Basin and Urban Area Regeneration

This initiative generates a plan to revitalize the land and environment in primary urban areas within ten years.

(d) Chemical Substance Risk Management

This initiative will build a system to assess and manage the control of chemicals through communications among various sectors of society to ensure safety and security within ten years.

(e) Global Water Cycle

This initiative will propose the optimum water management scenario and measures in Asia through development of scientific knowledge to clarify the projection and assessment of global water cycling changes. It will also provide a technological infrastructure for establishing water control technologies for sustainable global development.

4. GLOBAL WARMING RESEARCH INITIATIVE

Fundamental Recognition and Approach

International and domestic efforts must be made to resolve the problems of global warming defined in UNFCCC. A great responsibility has been imposed on Japan to meet the Kyoto Protocol adopted at the Third Conference of the Parties that signed the UNFCCC (COP3) held in 1997 in Kyoto. It is also important to extend international efforts, such as the Intergovernmental Panel on Climate Change (IPCC), to review the latest scientific knowledge regarding global warming. These themes require special commitment due to their great necessity and priority. IPCC started preparing the fourth assessment report in 2002 (to be completed in 2007), which includes discussions regarding the level to which greenhouse gases should be reduced. Japan is expected to contribute to the IPCC activities.

Wide-ranging efforts across various fields are required, since the causes and effects of global warming are closely related with social and economic structures. Approximately 80% of greenhouse gas emissions in Japan is CO_2 derived from energy consumption; thus, energy control technologies must be developed as part of the policy to develop technologies to minimize greenhouse gas emissions. Systematic and well-coordinated research is required, given the close relationships among phenomena such as global warming and climate changes, their effects, and countermeasures.

The Initiative first raised the key questions listed in Fig. 1; i.e., what scenarios of greenhouse gas emissions should be set to minimize increases in the concentration of greenhouse gases in the atmosphere that harm both humans and the global ecosystem. The Initiative also includes a hierarchical structure of a series of questions to resolve problems at individual levels.

The direction, theme, and scenarios for promoting R&D were developed based on a pyramid structure in which the roles, objectives, and relationships among them were clarified. Refinement of the question hierarchy resulted in programs for Monitoring and Process Study; Projection Modeling and Climate Change Study; Impact and Risk Assessment; Response Policies; Greenhouse Gas Fixation and Sequestration; and Anthropogenic Greenhouse Gas Emissions Reduction. R&D projects conducted by individual ministries will constitute one of these programs.

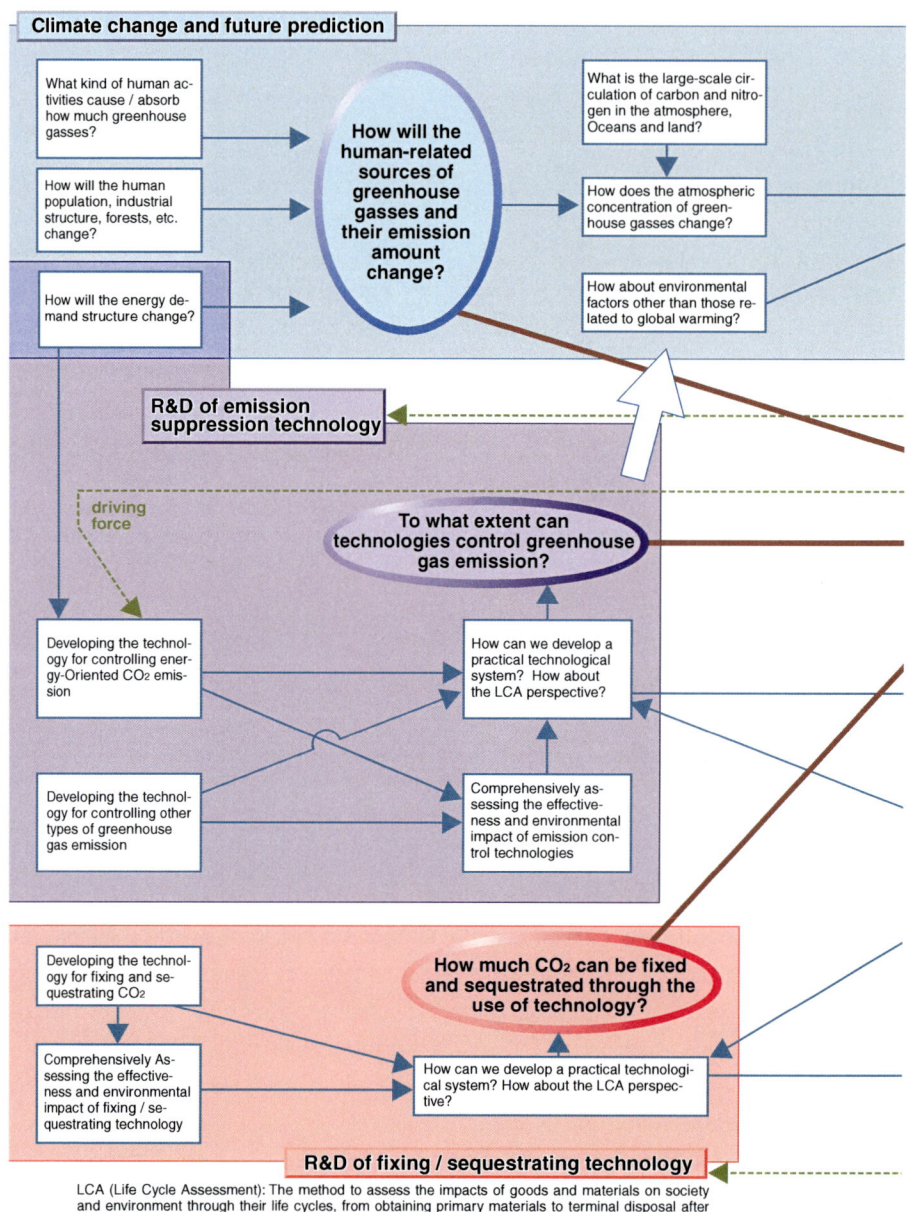

Figure 1. Key questions for resolving global warming.

Addressing the Problem of Global Warming

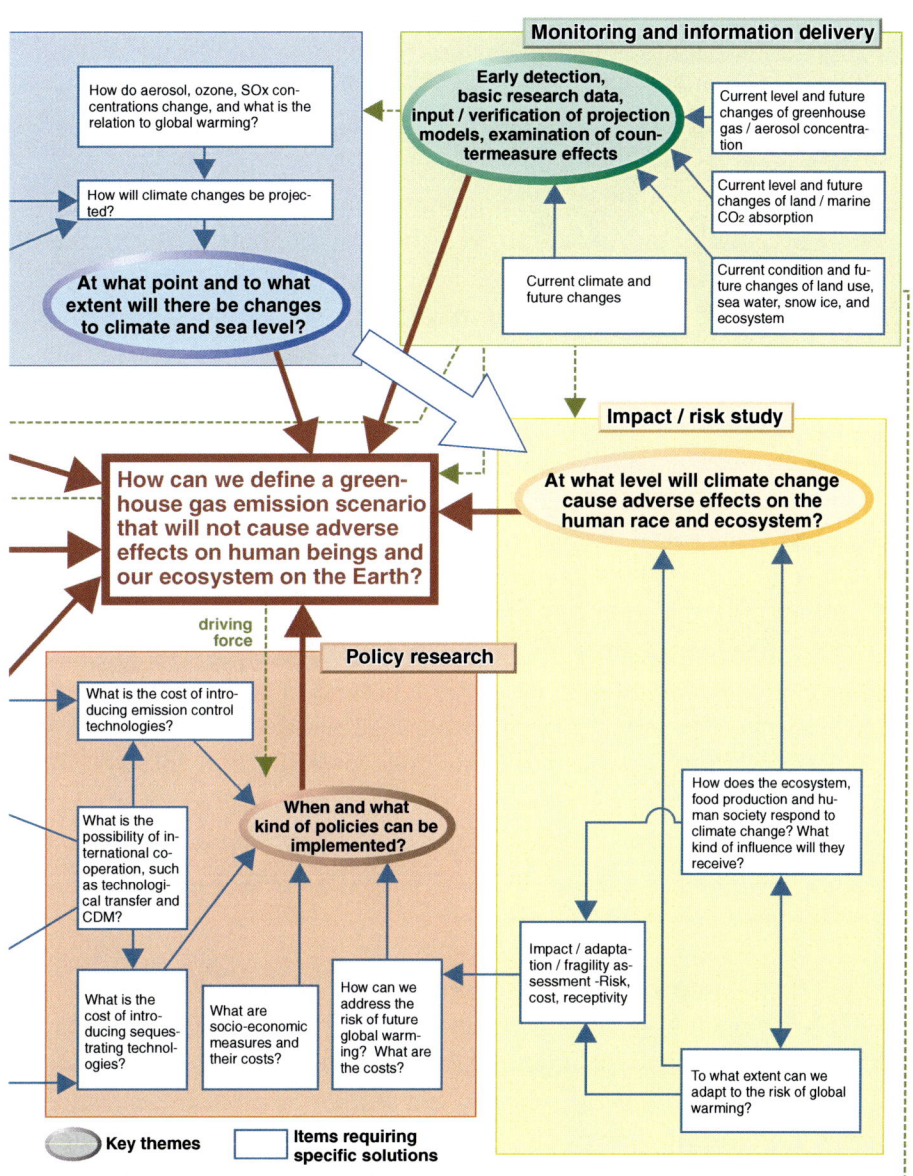

Program for Monitoring and Process Study

The immediate goal is to establish an integrated monitoring system of global warming focused primary on the Asia-Pacific region and data archives and delivery networks. The Ministry of Public Management, Home Affairs, Posts and Telecommunications; the Ministry of Economy, Trade and Industry; the Ministry of Land, Infrastructure and Transport; the Ministry of Agriculture, Forestry and Fisheries; and the Ministry of the Environment have promoted various monitoring projects of land, ocean and atmospheric components including AsiaFlux by using satellite, aircraft, ground stations, and ocean platforms. Further efforts are necessary to establish a structure responsible for archiving data from more expanded and improved monitoring, and to enhance the system for managing quality assurance and accuracy control of the observation data.

Program for Projection Modeling and Climate Change Study

The immediate goal is to attain the projection of climate changes due to global warming with higher precision by clarifying the mechanism of the global environment change and elaborating projection models for future greenhouse gas concentration and for climate changes. The ministries relevant to the Program are then required to promote the development of their original climate models with their own specialized bases and/or in co-ordination with one another and the implementation of projections.

Program for Impact and Risk Assessment

The immediate goal is to assess the overall influence of global warming and to propose appropriate countermeasures to avoid risks. The program will promote and implement joint research to assess the impacts of climate change on the ecosystem, disaster prevention/land conservation, water resources, food, forests, industry/energy, and health/civil life, based on the knowledge obtained from the two programs listed above. These efforts will more clearly specify vulnerable sectors and regions with greater risks of global warming. The final goal of this program is to explore the optimal comprehensive strategy for adaptation and mitigation of global warming.

Program for Response Policies

The immediate goal is to propose a countermeasure scenario to prevent global warming. The program will also include the development of a standard method for evaluation of the effectiveness of global warming mitigation technologies, estimation of future socioeconomic tendencies and the effects of relevant countermeasures to be taken, clarification of the adaptation and mitigation strategies, depending on the climate scenario, and establishment of international consensus-building techniques.

Program for Greenhouse Gas Fixation and Sequestration

The immediate goal is to develop technologies to reinforce the capacity of the ecosystem, including absorption of carbon dioxide by forests and technological innovations to separate, recover, fix, and reuse greenhouse gases from exhaust gases. These technology developments are required not only to fulfill the commitment to reduce greenhouse gas emissions within the first period (2008-2012) defined in the Kyoto Protocol, but also to conserve the future global environment.

Program for Anthropogenic Greenhouse Gas Emissions Reduction

The immediate goal is to develop technologies for energy-saving systems (e.g., solar generation) and new energy systems (e.g., solar generation, biomass energy and fuel cells) to reduce greenhouse gases, such as CO_2.

Organizational System for Implementing Initiatives

Two proposals were made during preparation of the initiatives. One was to provide CSTP members access to the front line of research. The other was to encourage productive discussions among the researchers who are participating in the initiatives, and thus provide a way for administrative officials to reflect their judgments in the policies. The Initiative Research Meeting (the liaison meeting of the relevant researchers) and the Liaison Meeting of Initiative-Promoting Ministries were therefore organized as illustrated in Fig. 2. In addition, the Environmental R&D Promotion

Project Team was established in April 2003 under the Expert Panel on Promotion Strategy of Prioritized Areas to realize better coordination among the five initiatives and to position the Initiative Research Meetings in official positions in CSTP. Each Initiative Research Meeting is regarded as an integral component of the Environmental R&D Promotion Project Team. The CSTP member responsible for the environment R&D ensures smooth implementation of the initiative by assessing the research plan and examining the progress status presented by individual ministries, through the Initiative Research Meeting and Environmental R&D Promotion Project Team meeting. The efforts of initiative activities, reported regularly to the Expert Panel on Promotion Strategy of Prioritized Areas, are used to revise strategies and to make policies for resource allocation, including budgets and human resources.

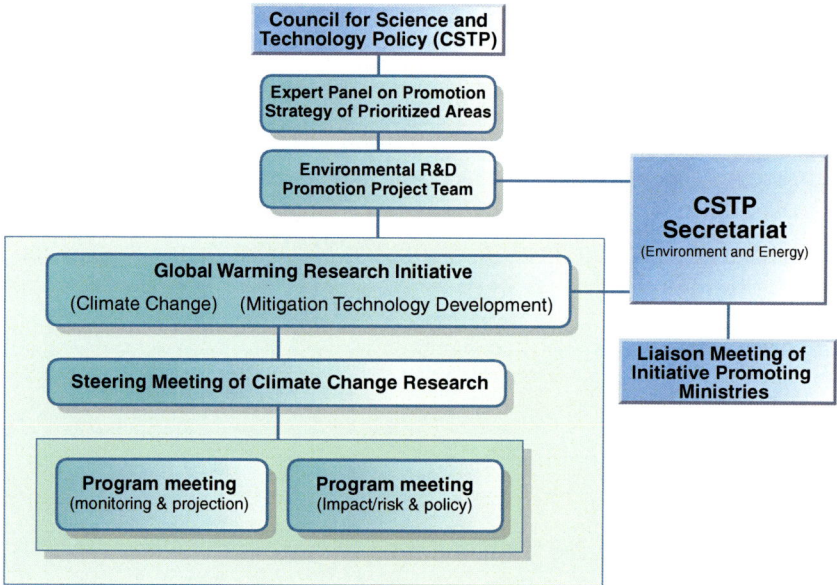

Figure 2. Initiative Promotion System.

Information sharing of execution plans, actual progress, and prominent results of the programs and projects are promoted in Initiative Research Meeting. The initiative members discuss policies to promote effective and efficient cooperation among the projects under the six programs and effective distribution of the primary results.

The research fields were integrated into two categories to efficiently promote the initiative. The first, "Climate Change Research Area", consists of four programs: the Monitoring and Process Study; Projection Modeling and Climate Change Study; Impact and Risk Assessment; and Response Policies. The second, "Mitigation Technology Development Area", consists of two programs, Greenhouse Gas Fixation and Sequestration and Anthropogenic Greenhouse Gas Emissions Reduction.

The Steering Meeting of Climate Change Research, attended by the chairman, program coordinators, and several researchers appointed by the chairman, was organized to discuss the common issues of the individual programs. In addition, two joint program meetings of monitoring and projecting model fields and influence/risk and policy research fields have been regularly held to encourage free and productive discussions among the relevant researchers from the institutions and the administrative officials from the ministries. These joint meetings promote effective cooperation among the projects and programs. The secretariats for the joint program meetings are organized in the Global Change Frontier Research System and National Institute for Environmental Studies (NIES).

The Liaison Meeting of Initiative-Promoting Ministries promotes the initiative based on administration and coordinates the policies in cooperation with the authorities connected with the ministries.

PART 2

TO WHAT EXTENT HAVE RESEARCHES ELUCIDATED GLOBAL WARMING?

CHAPTER 1

IS GLOBAL WARMING REALLY OCCURRING?
- WHAT GLOBAL MONITORING CAN TELL US

1.1 Introduction

Serious environmental pollution associated with human activities was first identified in Japan during the reconstruction boom after World War II. Many examples were reported in the 1950s and 1960s, such as the notorious Minamata disease and Yokkaichi asthma. Japan came to be known at this time as a "pollution power" as well as an economic power.

The number of pollution-related occurrences has decreased considerably since then. Emission reduction and the elimination of pollutants were relatively effective in improving the situation since the damage caused by specific pollutants, such as mercury, was restricted to the areas around the source.

Global warming, however, has emerged as a new environmental problem, differing from previous situations in that it takes place on a global scale. The sheer enormity of this makes it difficult to comprehend the entire scope of the problem.

The IPCC Third Assessment Report (TAR): Climate Change 2001 cites clear signs of global warming. The average surface temperature rose by 0.6 °C during the 20th century (possibly the highest temperature rise in a single century over the last thousand years). Moreover, satellite data collected since the 1960s indicate a depletion of Earth's snow and ice coverage and an increase in heat storage of the ocean.

Current and previous studies of global climate reveal a broad spectrum of worldwide global warming effects. Moreover, many long-term monitoring results indicate that the atmospheric concentration of greenhouse gases, such as carbon dioxide and aerosols, which impact global warming, is steadily increasing as a result of human activities.

Detailed research has been conducted into how these changes in atmospheric components caused by human activities influence global warming using a concept of radiative forcing. This research also includes climatic factors that arise from natural causes, including the change in solar radiation during the last several decades and supply of aerosols from large-scale volcanic eruptions.

The IPCC Third Assessment Report concluded that global-warming phenomena monitored over the last 50 years were likely caused by human activities.

This chapter will first discuss the long-term trend in atmospheric and marine temperatures, which can be used as a direct proxy of global warming. Phenomena directly influenced by global warming, including size change of mountain glaciers and changes in sea levels, are also examined. This discussion is followed by evidence that indicates that the temperature rise that occurred in the last half of the 20th century was a result of human activities.

Finally, we present a summary of monitoring methods and their results to reveal how changes in greenhouse gases, which may trigger the process of global warming, have been monitored. We also discuss how the carbon cycle, an important earth system effecting greenhouse gas emission, can be maintained and modified by anthropogenic processes.

1.2 How Has the Earth's Climate Changed?

1.2.1 What Determines the Earth's Surface Temperature?

The Earth's climate, comprised of phenomena such as atmospheric temperature and rainfall, is basically determined by the balance between solar radiation falling on Earth and the radiant energy released from Earth into space.

A variety of external factors, such as nature and humans, also exert influence. The concept of radiative forcing is useful for understanding these influences. For example, changes in solar radiation or large-scale volcanic eruptions are external factors of natural origin that can be interpreted as changes in radiative forcing.

A volcanic eruption creates negative radiative forcing that cools the Earth's surface by shielding it from solar radiation, but this effect lasts only several years. In contrast, human-induced greenhouse gases such as CO_2, CH_4, and N_2O produce positive radiative forcing that heats the Earth's surface. Increased atmospheric concentrations of these gases lead to global warming.

Therefore, it is important to monitor changes in atmospheric components, the sources of radiative forcing, and to quantify the strength of radiative forcing for each individual component.

The carbon cycle process, the key to understanding and predicting the amounts of CO_2 and CH_4 and other radiative forcing components in the atmosphere, must be studied quantitatively to predict prospective future atmospheric concentrations of greenhouse gases.

1.2.2 Changes in Atmospheric Temperature and Climate on Land

The IPCC Third Assessment Report describes changes in the surface temperature in the Northern Hemisphere over the last 1000 years based on data obtained from tree growth rings, coral, and ice core samples (Fig. 1).

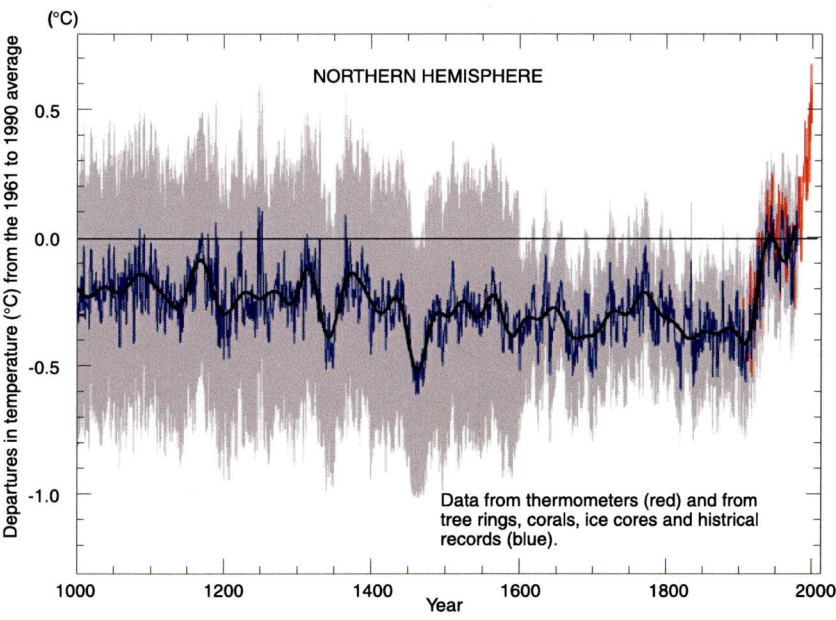

Figure 1. Variations of the Earth's surface temperature (Northern hemisphere) over the last 1000 years (IPCC, 2001). Blue curve and black curve indicate year by year and 50 year average variations, respectively, and the 95% confidence range in the annual data is represented by the grey region. Red curve indicates the data from thermometers.

The graph depicts the temperature changes expressed in terms of deviation from the annual averages between 1961 and 1990. It indicates that after tending to fall for 900 years before the 20th century, the temperature has started to rise dramatically.

A similar rise has been observed in Japan. The Japan Meteorological Agency monitored the annual average surface temperature in the 20th century at 17 observation sites where human impact on temperature changes as a result of urbanization is minimal (Fig. 2).

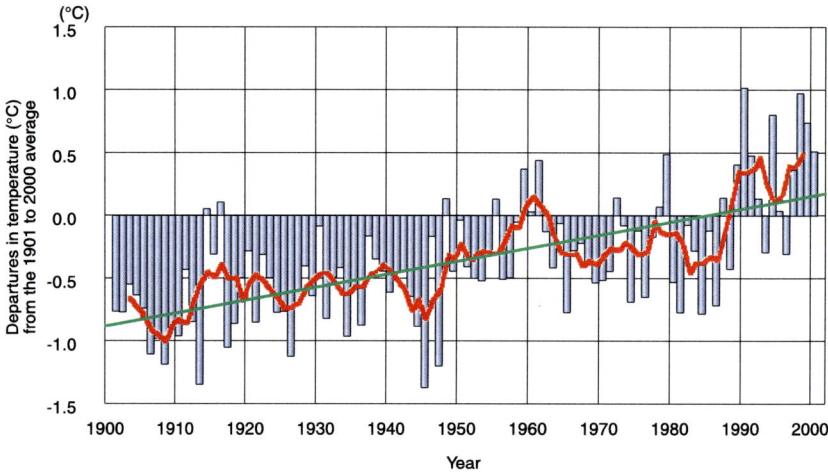

Figure 2. Variations of surface temperature in Japan from 1901 to 2000 (Japan Meteorological Agency, 2002). Bar graph indicates departures of annual average temperature from the 1901 to 2000 average, and red curve indicates 5 year running mean. Green curve indicates long term trend.

The temperature changed within the lower levels in Japan until 1940; it turned sharply upward in the 1960s and 1990s. The increase in the 20th century was approximately 1.0°C, which is much higher than the global average rate of about 0.6°C.

Middle-latitude countries like Japan are thought to be vulnerable to warming due to a decrease in the sun's reflection caused by reduced snowfall in high-latitude continents.

1.2.3 How the Ocean Temperature Changes

Oceans cover 70% of the Earth's surface and to a great extent govern the terrestrial climate by exchanging heat and moisture with the atmosphere.

Both oceans and atmosphere are categorized as fluids, but oceans can store 1000 times as much heat as the atmosphere, and therefore ocean temperatures rise very slowly. Vertical mixing processes, through which heat is conveyed, are also slow in the ocean. This slow vertical mixing suggests that ocean surface waters respond to atmospheric temperature changes somewhat quickly, while the middle and deep layers in the ocean may take more than several years to be effected, after which they exert greater long-term impact.

Temperatures of the upper ocean in different areas on the same latitude vary by seasons, but there is also a recurrent pattern of changes over the years.

For example, El Niño and La Niña, which cause large-scale water temperature changes in the equatorial Pacific region, follow a clearly recognizable pattern of cycles that recur every several years.

A variety of factors that determine the climate system are thought to be responsible for creating the above cycles. However, their frequency and duration have increased compared to 100 years ago, and the degree of change has intensified.

There have not been many studies reporting the results of long-term monitoring of ocean temperature; however, the findings of one such survey, in offshore Bermuda from 1922 to 1995, have been published (Joyce and Robbins, 1996). That survey indicates a clear long-term tendency of temperature increase (0.47°C over 100 years) in this subtropical deep-sea (1500 to 2500m) water. This rate is equivalent to the temperature increase necessary to raise sea level by 7cm over 100 years by heat expansion.

In contrast, monitoring results from deep-sea water in a sub-arctic zone of the Atlantic have indicated a fall in temperature, which is the direct opposite of the Bermuda report. Thus, it remains unclear whether the Bermuda result reflects warming of the atmosphere. A research project analyzing monitoring data of marine temperatures over the last 50 years from waters across the globe (at depths ranging from surface levels to 3000m) indicates that the temperature has risen by approximately 0.05°C.

These observations suggest the possible impact of global warming on the ocean, which functions as a vast heat storage facility in the climate system, when the temperature of the Earth's surface rises. However, these observations are not sufficient to verify that the same warming tendency found in terrestrial temperature exists in the ocean. Additional long-term observations are required to confirm the trend.

The Japan Meteorology Agency has released data that verifies surface water temperature anomalies in winter (December to February) in the central North Pacific over the last 100 years (Fig. 3).

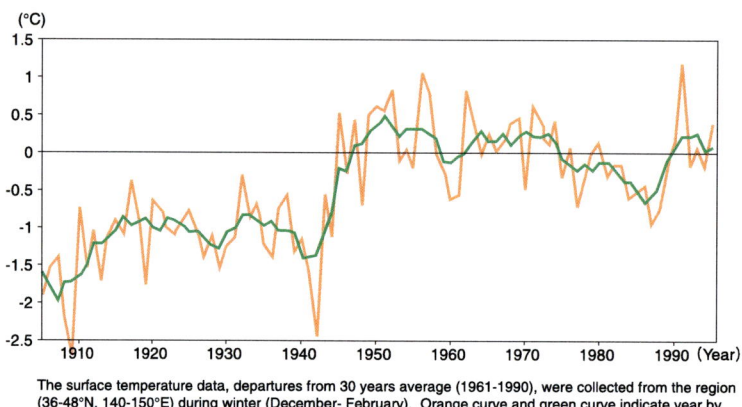

Figure 3. Variations of winter sea surface temperature in the Central North Pacific (1905-1995) (Japan Meteorological Agency, 2002).

Figure 3 combines data collected by the Kobe Marine Observatory since the early 20th century (data obtained by merchant ships, fishing boats, observation ships, and the former Japanese Navy in the central Northern Pacific) and figures published by the Comprehensive Ocean-Atmosphere Data Set (COADS). These data are important to understanding long-term climate changes. The figure indicates a dramatic rise of more than 1°C in the water temperature in the central Northern Pacific during the 1940s.

1.2.4 Have Glaciers and Oceanic Ice been Effected by Global Warming?

Monitoring data from mountain glaciers as well as that from polar ice sheets and oceanic ice floes are prominent among the data often used to confirm the effect of global warming. About 2.1% of the global water reservoir is in the form of glaciers. Therefore, glacier conditions correlate directly with sea-level changes as well as being an index for climate change. Monitoring reports from satellite observations of sea ice in the Arctic Sea reveal that the volume of ice in summer has been reduced by 10 to 15% compared with the level in the 1950s. This reportedly corresponds with an atmospheric temperature rise in spring in the high latitude zones of the

Northern Hemisphere.

Recent observations by US and Russian submarines revealed a clear decrease in the thickness of the Arctic ice over the last several decades.

Mountain glaciers have been disappearing over the last several decades, but these are limited in area. Shrinkage of the Antarctic ice sheet has not yet been reported.

There is a confirmed reduction in the thickness of the snowy ravine at Tsurugidake, Toyoma Prefecture in Japan; a connection with global warming is assumed.

The documenting of Lake Suwa in Nagano Prefecture represents one of the best historical records of climate change in the world. The occurrence of a unique phenomenon known as *Omiwatari*, "the divinity's pathway," was observed at Lake Suwa in winter. The icy surface swells to make a long ridge if the temperature remains very low after the lake's surface is completely frozen. This natural process has been incorporated into a religious rite of the Suwa Shrine, and records have been kept since 1443. This information is very important to understanding how the winter climate has changed over the last several centuries.

The combined data from instrumental observation after 1898 and the estimated temperatures prior to that year by inference from the data when "*Omiwatari*" occured are provided in Fig. 4.

Omiwatari has not been seen very often in recent years, presumably due to a series of warm winters. However, this is unconfirmed; water pollution also may exert an impact.

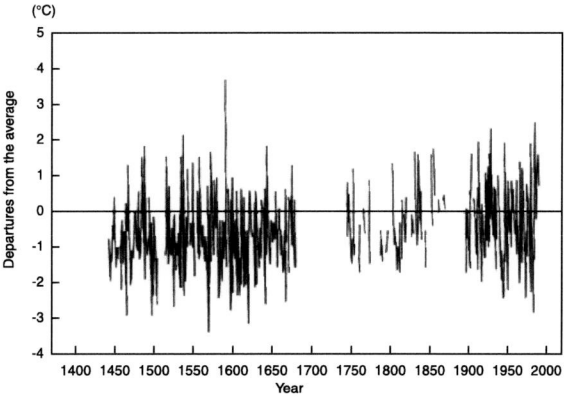

Figure 4. Variations of winter (December and January) temperature of Suwa region, Nagano Prefecture, Japan, over the last 500 years estimated from ice condition of the Lake Suwa (Mikami and Ishiguro, 1998).

An increase in greenhouse gases is likely to affect the Earth's water circulation.

There have been many natural disasters in recent years, with extreme rainfall causing floods due to tropical low pressure in addition to contrasting periods of drought. These events have received considerable attention.

However, the two large databases presently available substantiate no increase in worldwide rainfall over the last 20 years. They instead reveal the reverse tendency. This result conflicts with predictions of many global warming models that indicate that global warming by an increase in greenhouse gases would activate global water circulation, and thus increase rainfall.

1.3 Atmospheric Concentration of Greenhouse Gases and Aerosols: Influence on Past, Present, and Future Climate

Global warming is caused when greenhouse gases such as CO_2 are produced through human activities, such as burning fossil fuels and deforestation, and remain in the atmosphere. Long-term monitoring of greenhouse gases on a global scale is vital to identify the sink-source processes, to predict future conditions, and to determine appropriate countermeasures.

The most important greenhouse gas is water vapor in the air. However, this is normally treated separately since its concentration is determined by the climate system regardless of human influence. The next most important gas is CO_2. Figure 5 illustrates the proportion of greenhouse gases in an enhanced greenhouse effect.

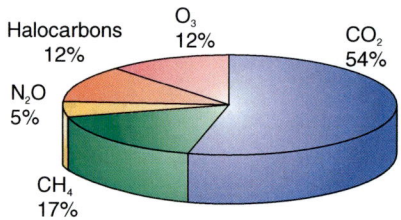

Figure 5. Radiative forcing due to changes in greenhouse gases from pre-industrial time to the present.

A reduction of CO_2 emissions from human activities and increased capacity for absorption and repair in the atmosphere are required to suppress the increase of CO_2 in the atmosphere to an acceptable level and to stabilize this concentration. The 3rd Session of the Conference of the Parties of the United Nations Framework Convention on Climate Change (COP3) in Kyoto, held in December 1997, recognized the urgent need to establish appropriate management of the forest ecosystem by activities such as afforestation and reforestation to increase absorption and repair.

1.3.1 An Archived Atmosphere over Hundreds of Millennia in the Antarctic Ice Core

Only 45 years have passed since humans began to monitor CO_2 concentrations in the atmosphere. Ice from the Antarctic or Greenland can be used to obtain data for earlier periods by boring ice samples and analyzing the atmospheric layers trapped in the snow that have been compressed as hard ice. This monitoring is conducted in Japan by the National Institute of Polar Research, in conjunction with Tohoku University and other institutions.

Figure 6 depicts changes in the CO_2 concentrations and temperature for the past 320,000 years. Temperatures can be estimated by the oxygen isotope ratio (^{18}O) in water molecules in the ice. This graph reveals that a warm period with high CO_2 concentrations appears roughly every ten millennia. Scientists continue to argue over cause and effect, whether an increase in CO_2 led to global warming or a temperature rise led to an increase in CO_2.

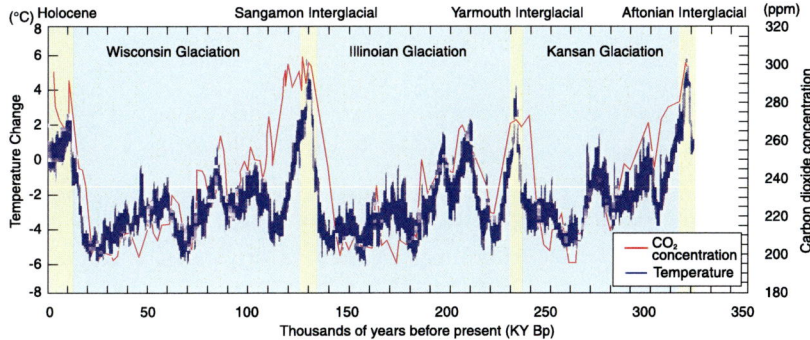

Figure 6. CO_2 concentrations over the past 320,000 years from 250m-thick ice core records from Dome Fuji in Antarctica. Temperature is estimated from the oxygen isotope ratio of ice water (based on Kawamura *et al.*, 2003).

1.3.2 Results of Recent Direct Monitoring: Correlation Between Human-Induced Emissions and Climate Change

Dr. Keeling started to monitor CO_2 concentrations in Mauna Loa in Hawaii and the South Pole in 1958. These results exhibit continuity with the data from Antarctic ice analyses (Fig. 7). The CO_2 concentration before the Industrial Revolution, when human influence was minimal, is estimated to be about 280ppm.

There are presently 30 sites in the world where regular monitoring is conducted, including several sites under Japanese supervision, such as Ryori, Minamitorishima, Yonakunijima (Japan Meteorology Agency); Hateruma Island, Ochiishi (National Institute for Environmental Studies); and Syowa Station and Svalbard (National Institute of Polar Research and Tohoku Univ.). There is no direct impact from human activities or forests at these sites; they can therefore be treated as background. There are also 100 sampling points where samples have been taken for analysis.

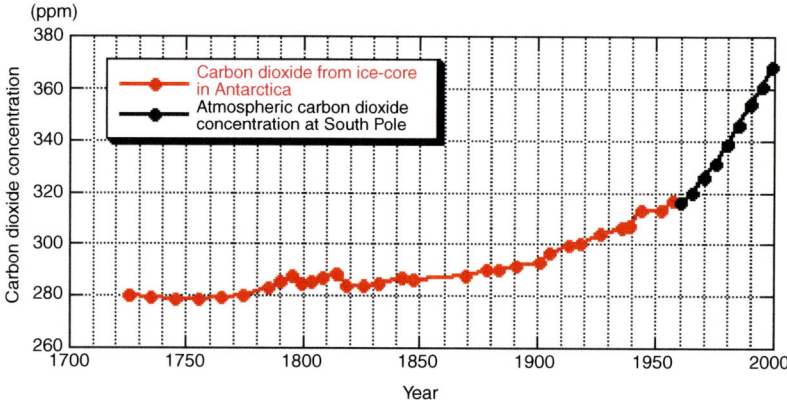

Figure 7. Variation of CO_2 concentration over the past 250 years. The record before 1953 is from the ice core (Machida, 1990) and that after is from the South Pole observatory.

Figure 8 displays the monitoring results from Americans at Mauna Loa by NOAA and Japanese at Minamitorishima, Hateruma Island, and Syowa Station. All except for Syowa Station reveal an increase while repeating similar seasonal changes. The data from Syowa Station exhibit very small seasonal variations. The seasonal changes indicate that the concentration is actually greatest in spring. The concentration levels then decrease due to the forest absorption of CO_2 through photosynthetic reaction until the beginning of autumn, when they start to increase due to the release of CO_2 through respiration by soil and plants; the forests absorb CO_2 through photosynthetic reaction. The atmosphere is relatively well mixed by air circulation at any given latitude, and the seasonal variations are repeated with the same amplitude at each latitude. The high-latitude zone in the Northern Hemisphere has a greater amplitude due to its large proportion of forest area; the Antarctic, where there is no forest, exhibits extremely small seasonal changes.

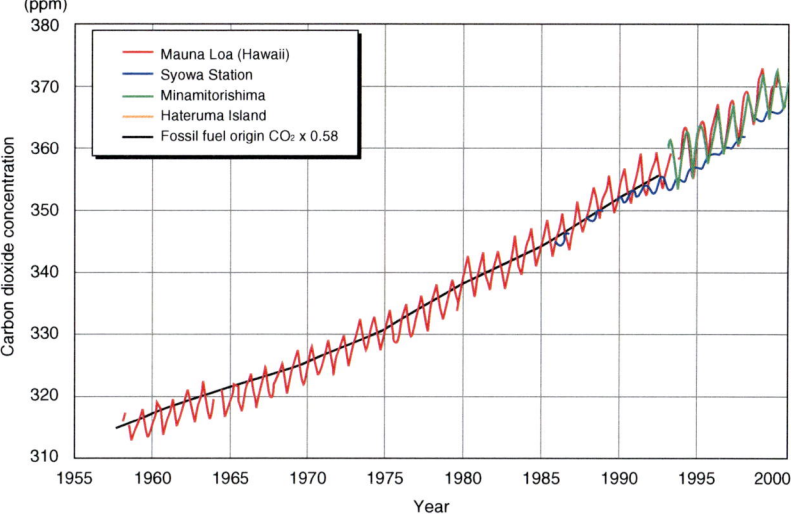

Figure 8. Long-term carbon dioxide monitoring data at Mauna Loa (NOAA), Syowa Station (National Institute of Polar Research and Tohoku University), Minamitorishima (Japan Meteorology Agency), and Hateruma Island (National Institute for Environmental Studies). Black line indicates the 58% value of accumulated carbon dioxide from fossil fuel combustion.

The long-term increase in concentration can be isolated by excluding seasonal variations; they can be factored out because of their regularity. A remarkable similarity is apparent when the resulting figure is compared to the accumulated emissions from fossil fuel, at 58%. It is therefore clear that the long-term CO_2 increase is a result of human activities, and that about 60% of this is accumulated in the atmosphere.

It is evident that this long-term increase rate is not constant. Figure 9 shows the CO_2 increase rate in the tropics as well as indices for El Niño / La Niña, and for sea-surface and atmospheric temperatures. The figure reveals a close relationship between the CO_2 increase rate and temperature deviation from the normal year level, except for the early 1990s, when the impact of the eruption of Mt. Pinatubo was strong. This finding suggests that global warming will accelerate the increase in CO_2 concentrations.

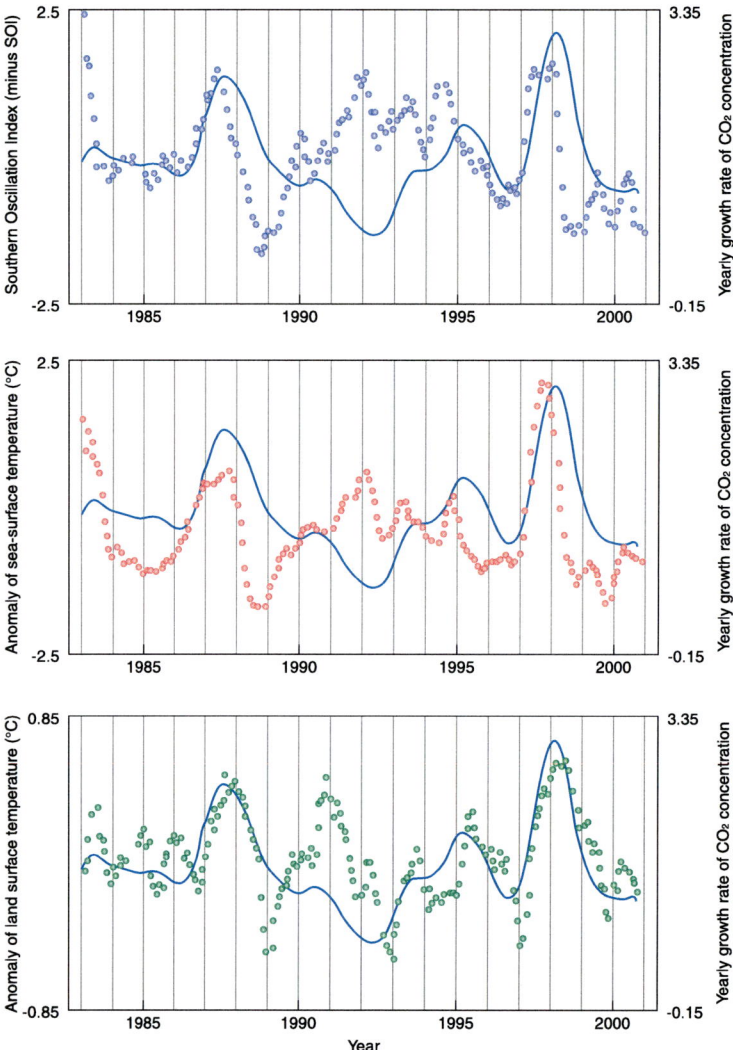

Figure 9. The growth rate of CO_2 concentration in the tropical zone (30°N-30°S) and Southern Oscillation Index (SOI) (upper panel); anomaly of the sea surface temperature (middle panel); anomaly of land-surface temperature (lower panel). The land-surface temperature anomaly is from NCEP reanalysis data (1000hPa). Curves are CO_2 growth rate, circles are five-month running averages of each element (produced by Japan Meteorology Agency from WMO WDCGG database).

1.3.3 Methane, Nitrous Oxide, CFCs, Ozone, and Other Greenhouse Gases

Methane is generated from various sources (Fig. 10), but the rate of accumulation in the atmosphere is decreasing (Fig. 11). One possible reason for this is that the increased methane concentration has accelerated the atmospheric process by which it is removed (reaction between OH radical and methane), and stabilized the atmospheric concentration. Nitrous oxide is stable in the troposphere and only disappears when it reaches the stratosphere by photochemical reaction. It stays in the atmosphere for more than 110 years. The amount generated is minimal but is steadily increasing. Halogenated hydrocarbon production (CFCs) (Fig. 12) was totally banned after discovery of the ozone hole to protect the stratosphere ozone layer. The ban has prevented any increase in the atmosphere since the early 1990s. Although they stay in the atmosphere for a very long time, 45 years for CFC-11 and 640 years for CFC-13, the atmospheric concentration has begun to decrease. This illustrates that the concentration of greenhouse gases in the atmosphere will definitely decrease if we commit ourselves to its reduction. However, unlike CFCs, emissions of some gases, such as CO_2, are closely connected to the scientific and technological basis of modern civilization, which makes it difficult to substantially reduce their emissions.

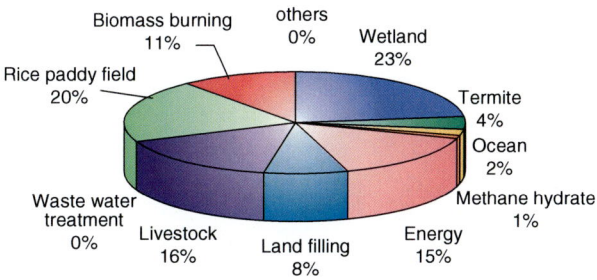

Figure 10. Source strength of Methane.

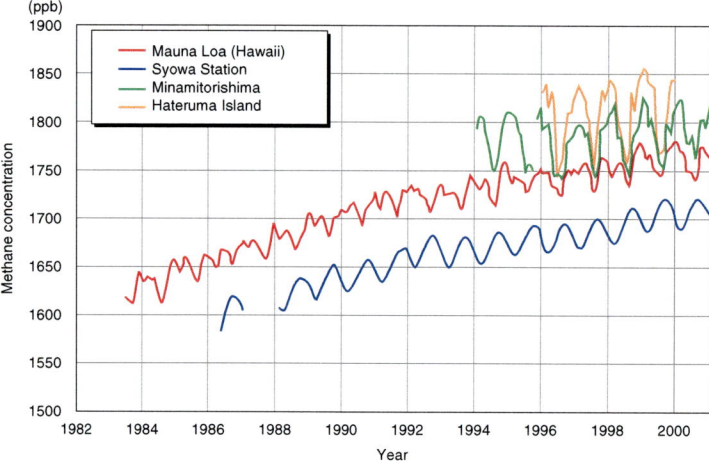

Figure 11. Long-term methane monitoring data at Mauna Loa (NOAA), Syowa Station (National Institute of Polar Research and Tohoku University), Minamitorishima (Japan Meteorology Agency), and Hateruma Island (National Institute for Environmental Studies) (from WMO WDCGG database).

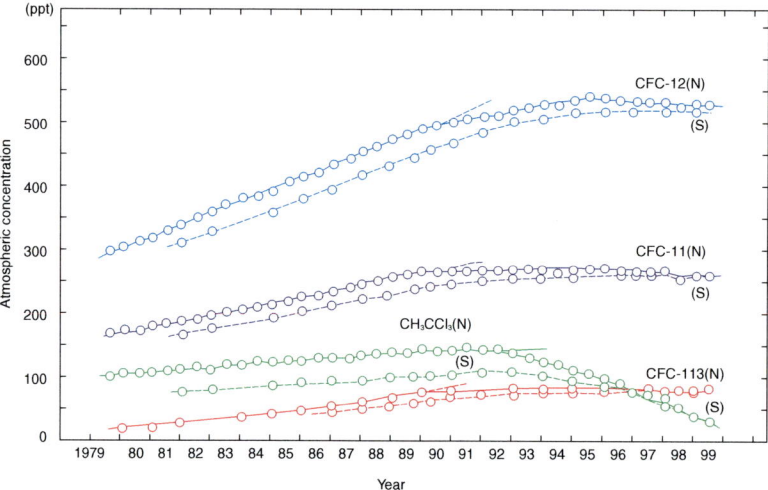

Figure 12. Trend of primary halogenated hydrocarbons in the atmosphere (by Tokyo University, N: Hokkaido, S: Syowa Station) (Ministry of the Environment, 2001).

While nitrogen oxides emissions are increasing through the burning of fossil fuel, the amount of ozone in the troposphere is also increasing. Moreover, burning fossil fuel also generates gases like sulfur dioxide (SO_2), which leads to sulfuric acid mist formation. Soot from combustion also increases the aerosols in the air. Sulfuric mist is thought to have a cooling effect by scattering solar radiation and accelerating cloud generation, which acts as a shield against sunshine. Aerosols are thought to have a warming effect because they absorb sun's heat. However, there is currently no evidence in the monitoring data that indicates an increase of aerosols. They stay in the air for only limited periods, unlike long-lasting greenhouse gases, but under normal circumstances can have a significant impact on climate.

1.4 Terrestrial and Oceanic Sources and Sinks for Major Greenhouse Gases

1.4.1 Sink Ratios for Carbon Dioxide on Land and in the Oceans

Understanding in detail the various mechanisms and different time scales by which carbons circulate among fossil fuels, the atmosphere, forests, and oceans is vital to forecasting future changes in the atmospheric concentration of CO_2 (Fig. 13).

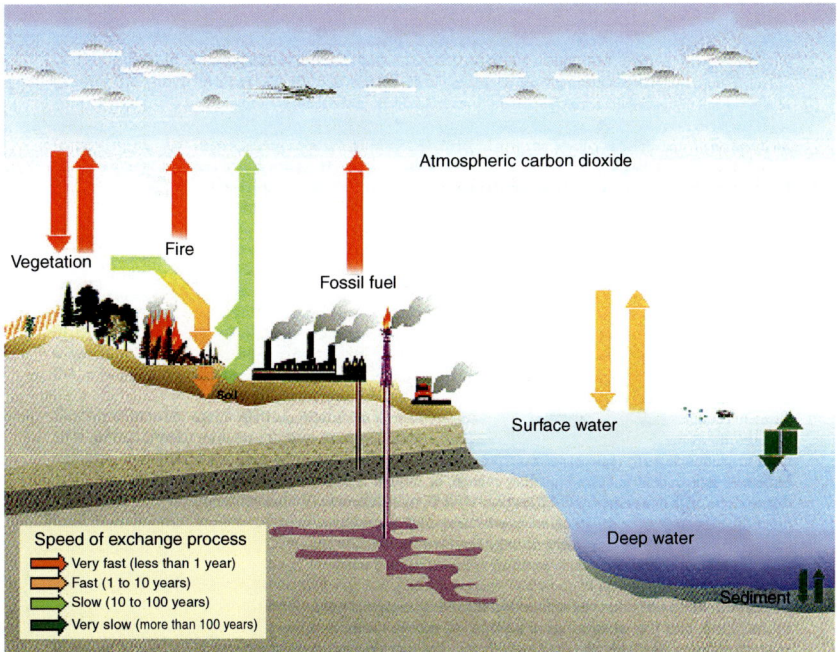

Figure 13. Carbon cycle processes and their time scales (after IPCC, 2001).

Relatively rapid circulation and accumulation of CO_2 occurs in forests and soil in terrestrial areas, and it is also absorbed into sea-surface waters. Forty percent of the total amount of CO_2 on a 10-year basis consists of fossil fuel consumption, but the ratio varies considerably year to year. The annual CO_2 absorption amount ranges from 0.5Pg to 4.5Pg. The variations are created by temperature changes connected to phenomena like El Niño. Carbon cycle models suggest that a rise in temperature leads to an increase in CO_2 emissions from the soil.

The CO_2 in the atmosphere includes two types of carbon. Ninety-nine percent is ^{12}C, with a mass number of 12, and 1% is its isotope, ^{13}C, with a mass number of 13. Fossil fuels, which are derived from prehistoric plants and animals, contain less ^{13}C since plants absorb ^{12}C more effectively. Therefore, fossil fuel use decreases the ratio of ^{13}C in the air. However, when the organisms living on land absorb CO_2, the ratio of ^{13}C increases in the atmosphere. The absorption of CO_2 in oceans is primarily a physical process, and the same rate applies to both isotopes. The ratio of terrestrial to oceanic CO_2 absorption can be obtained using these factors (Fig. 14).

This also applies to changes in oxygen concentration. Earth's oxygen supply comes from the accumulation of oxygen produced by plants through photosynthetic reaction. This oxygen is consumed when fossil fuel is burned. Plants can replace oxygen loss as a result of CO_2 repair through photosynthesis as they absorb CO_2 and produce oxygen. Absorption into the ocean has no impact on the oxygen concentration.

Is Global Warming Really Occurring? - What Global Monitoring Can Tell Us 35

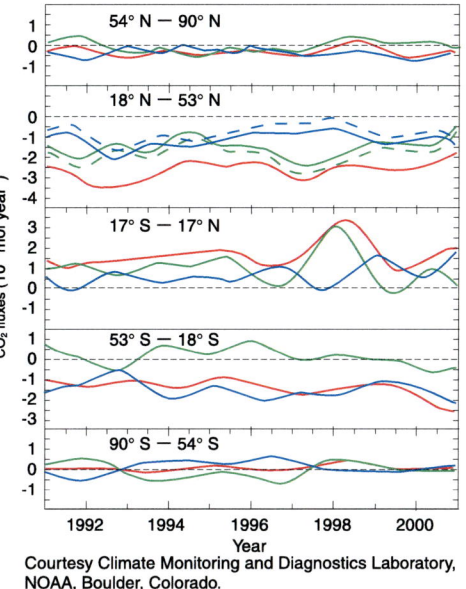

Figure 14. Reconstructions of net CO_2 fluxes between the atmosphere and the terrestrial biosphere (green) and the ocean (blue) as well as the total net flux between the atmosphere and the Earth's surface from pCO_2 (red), in major latitude zones. Dashed line is neutral, plus is emission to the atmosphere (Tans *et al.*, 2001).

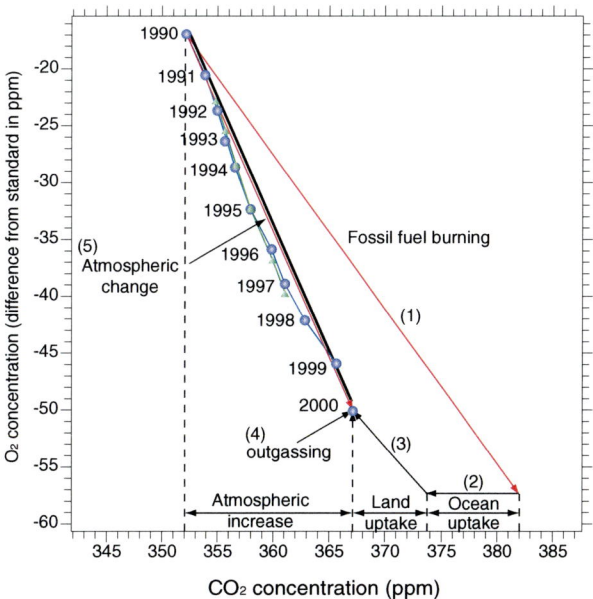

Figure 15. Partitioning of fossil fuel CO_2 uptake using O_2 measurements. The graph shows the relationship between changes in CO_2 (horizontal axis) and O_2 (vertical axis) (IPCC, 2001).

Arrow 1 in Fig. 15 indicates the oxygen decrease and CO_2 increase from 1990 to 2000 by fossil fuel consumption; arrow 2 indicates the CO_2 decrease by the ocean absorption and the oxygen remaining constant; arrow 3 indicates the CO_2 decrease and oxygen increase mediated by terrestrial ecosystem; the very small arrow 4 designates the oxygen released from seawater due to water temperature rise. Arrow 5 should match the figure for 2000 on the line with circles, indicating the yearly CO_2 increase and oxygen decrease. The absorption rates for the oceans and continental areas can be estimated using known data for the levels of fossil fuel consumption and seawater temperature increase and by matching two arrows. The result is that oceanic absorption is 1.7±0.7PgC/year and terrestrial absorption is 1.4±0.7PgC/year.

A precision of seven orders is required to monitor changes in oxygen concentration. An automatic continuous monitoring method using gas chromatography was developed in Japan, and monitoring has been conducted at the Hateruma Monitoring Station (Fig. 16).

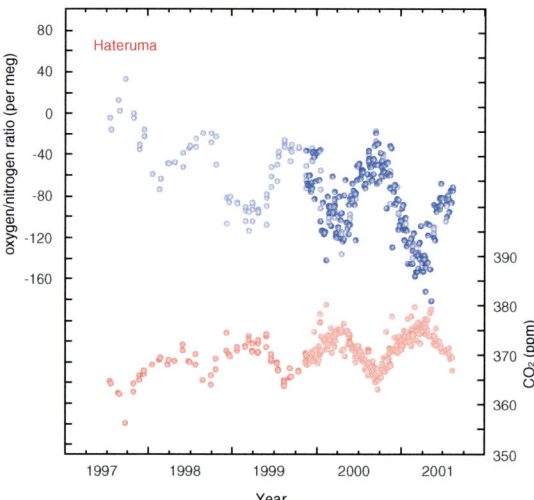

Figure 16. Atmospheric concentration of carbon dioxide (red circles) and oxygen/nitrogen ratio (blue circles) observed at Hateruma monitoring station by National Institute for Environmental Studies (Tohjima, 2000).

per meg
Unit of oxygen/nitrogen ratio in the atmosphere defined as

$$\left(\frac{[O_2/N_2]_{\text{measurement}}}{[O_2/N_2]_{\text{standard}}} - 1\right) \times 10^6$$

1ppm corresponds to 4.8 per meg

1.4.2 How Terrestrial Carbon Dioxide Absorption Works

Carbon on the Earth exists in different regions, including the atmosphere, oceans, terrestrial vegetation, and soil; it circulates among these regions while changing its form among gases, inorganic carbon, and organic carbon (a process which is known as the carbon cycle). A CO_2 increase in the air indicates that the amount of carbon stored in the atmosphere is increasing because the CO_2 released into the air (into atmospheric storage) by human activities, such as fossil fuel use and deforestation, has exceeded the amount of absorption and sequestration by the oceans and by terrestrial vegetation stores. It is essential to clarify the quantitative movements of carbon in each region to understand the control factor of atmospheric CO_2 concentration and its future trend.

Carbon Circulation and Sequestration in the Forest Ecosystem

The fact that forests absorb atmospheric CO_2 to produce organic matter through photosynthetic reaction is well known. They use most of the products for respiration, and return CO_2 to the atmosphere. Only the residue can be counted as an increase in forest plants. When living parts of forests die, they fall to the ground as leaves and branches. The surface soil contains substantial organic matter, the quantity of which depends on how much organic matter fallen leaves and branches bring and on how much CO_2 is released from the decomposition of organic matter by microbes. The speed of organic decomposition is dependent on soil temperature and moisture.

Figure 17 illustrates the quantitative relationships that exist in a forest ecosystem. Many elements are correlated in the processes of carbon absorption and decomposition in the forest ecosystem among both living forest plants and soil. However, the precise mechanisms of the carbon cycle in the forest ecosystem have yet to be revealed. The following problems must first be resolved.

The changes in carbon accumulation both in stands and in soil (as organic carbon) must be monitored on a long-term basis, and the results correlated with those for CO_2 exchanges between forest ecosystems and the atmosphere.

Quantitative studies, conducted in open-air environments, of the acceleration in forest photosynthetic reactions associated with the annual increase in atmospheric CO_2 concentrations are required. This CO_2

fertilization effect has already yielded demonstrable benefits for greenhouse agriculture.

We must determine whether there is any positive effect on forest growth from global warming, or from soil eutrophication due to nitrate and ammonium salt in the air that fall on the ground and permeate the soil with rain. The effects of acceleration in organic decompositoin due to rising temperatures also must be studied.

These subjects require long-term monitoring and research conducted at many different ecosystems. This will certainly not be easy to achieve. The basic data and knowledge currently available to us is far from sufficient to resolve the above issues.

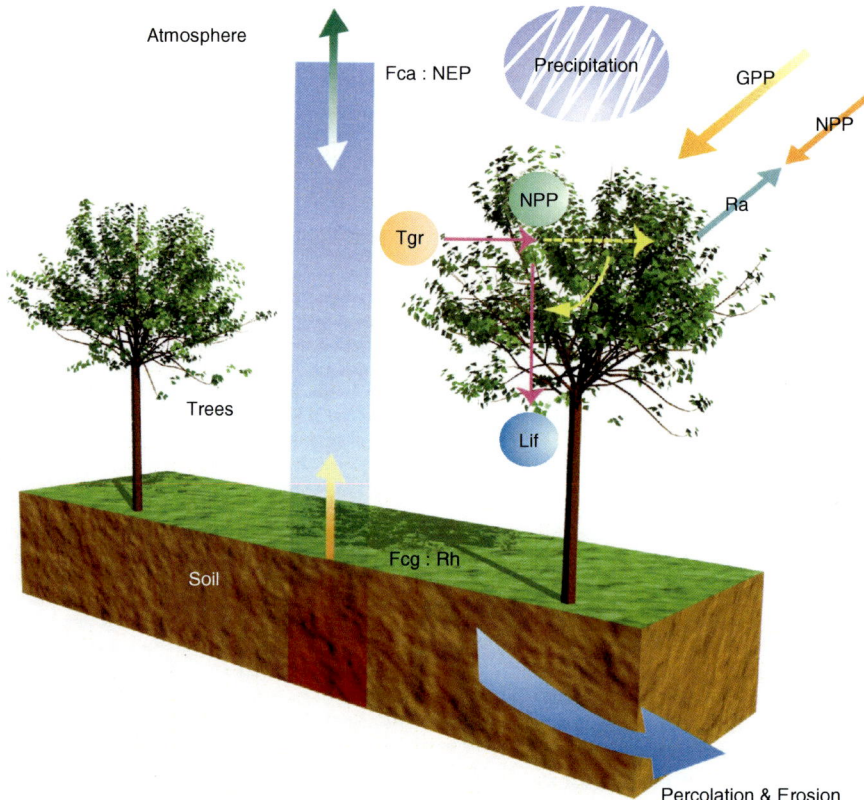

Figure 17. Carbon flow and stock in forest ecosystem.
GPP: Gross Primary Production (Photosynthesis), NPP: Net Primary Production, NEP: Net Ecosystem Production, Ra: Plant respiration, Rh: Soil respiration, Tgr: Plant biomass increase, Lif: Fallen leaves and dead branches.

How Carbon Dioxide Absorbed by Forest Ecosystems can be Measured

The CO_2 flux between the atmosphere and the forest ecosystem (Fig. 17) is measured on a tower over a forest by measuring the covariance of CO_2 concentration and the vertical wind velocity. The relationship between this CO_2 flux and climatic conditions, along with any seasonal and yearly changes in the carbon balance, are then analyzed. Observations using a tower are essential to develop carbon cycle models for forest ecosystems, since they enable us to learn directly about the CO_2 flux between the upper part of a forest and the atmosphere at a specific point. A tower is also suitable for long-term continuous observations.

It is also important to conduct parallel research in the forest ecology in the area, covering such aspects as forest photosynthetic reactions; respiration; soil decomposition; and the height, diameter and growth of trees. Several research institutes and universities throughout the world have already initiated tower flux observations. Research programs of the International Geosphere-Biosphere Programme (IGBP) have established long-term monitoring for a variety of plant ecosystems as a predominant theme in the core projects of Biospheric Aspects of the Hydrological Cycle (BAHC), International Global Atmospheric Chemistry Project (IGAC), and Global Change and Terrestrial Ecosystems (GCTE) across north-south or wet-dry transects.

The National Institute of Advanced Industrial Science and Technology (AIST), in conjunction with the River Basin Research Center, Gifu University, has been conducting a continuous tower observation for CO_2 flux and climatic conditions since October 1993. The flux of heat and vapor, climatic conditions including temperature, wind, humidity, and radiation, forest biomass, types of trees/height of canopy, Leaf Area Index (LAI) (total area of tree leaves lying over unit-area), soil moisture/temperature, soil respiration, and the amount of fallen leaves/dead branches and their rate of decomposition are monitored and studied, as well as the CO_2 flux between atmosphere and forest ecosystem, and the CO_2 concentration. These data are regarded as unique and invaluable long-term monitoring results, even on the global level.

Figure 18 illustrates the seasonal and year-to-year changes of CO_2 flux in deciduous broadleaf forest in the cool temperate zone (monthly totals for CO_2 exchange amounts for daytime, nighttime, and the overall 24-hour period from Oct. 1993 to Dec. 1998). The monitoring site is situated 1420m above sea level; the average temperature is 7.3°C; annual rainfall is 2400mm;

snowfall is 1 to 1.5m; and the predominant trees are *Betula platyphylla var. Japonica* and *Quercus crispula*, which are 15 to 20m high. The forest becomes active at the end of May with increased photosynthesis, and the CO_2 exchange amount becomes positive (i.e., CO_2 is absorbed into the forests). This amount peaks in July and August. The forest becomes less active in September and its CO_2 absorption rate decreases rapidly. The leaves fall in early October, and there is snow from early December to mid-April. The flux changes yearly; for example, the daytime data for summer 1994 is only 70% of that for summer 1995. This tendency is particularly conspicuous in June and July. These annual variations probably occur because solar radiation, temperature, and rainfall in the rainy season impact the forest's capacity for photosynthesis and decomposition of organic matter. This suggests that future climate changes will affect the CO_2 absorption ability of forest ecosystems.

Figure 19 depicts CO_2 absorption at different latitudes based on estimates from the results of long-term flux monitoring in Europe (a minus figure represents absorption from the atmosphere by forests). The graph indicates that forest ecosystems absorbed CO_2 at all the sites, even if the amount varied significantly (0 to 7tC/ha/year).

The monitoring results of East Asia and Europe indicate that the amount of carbon absorbed by the forest ecosystem is 300gC/ m^2 (3tC/ha) per year. This result leads us to estimate that 1.6 PgC/year of CO_2 is absorbed by protected forests throughout the world (5.3 million km^2, 13% of Earth's total forested area). Interestingly, on-site direct results of monitoring the CO_2 flux indicates that only protected forests achieved the level for CO_2 absorption/sequestration evaluated by IPCC (0.5~1.9PgC per year), as a result of forest recovery and similar activities (IPCC, 1996).

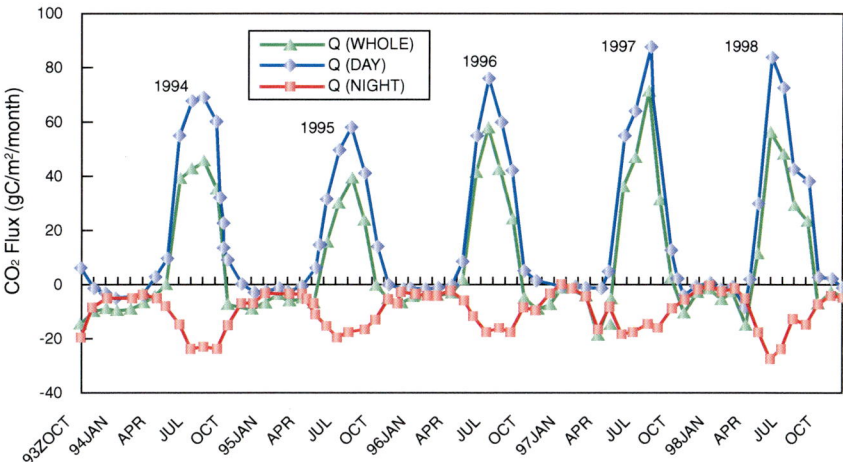

Figure 18. Seasonal and yearly variation of carbon dioxide flux observed over deciduous broadleaf tree at Takayama (whole day:Q(WHOLE), daytime :Q(DAY), nighttime :Q(NIGHT)) Plus is from the atmosphere to the forest (Yamamoto *et al.*, 1999).

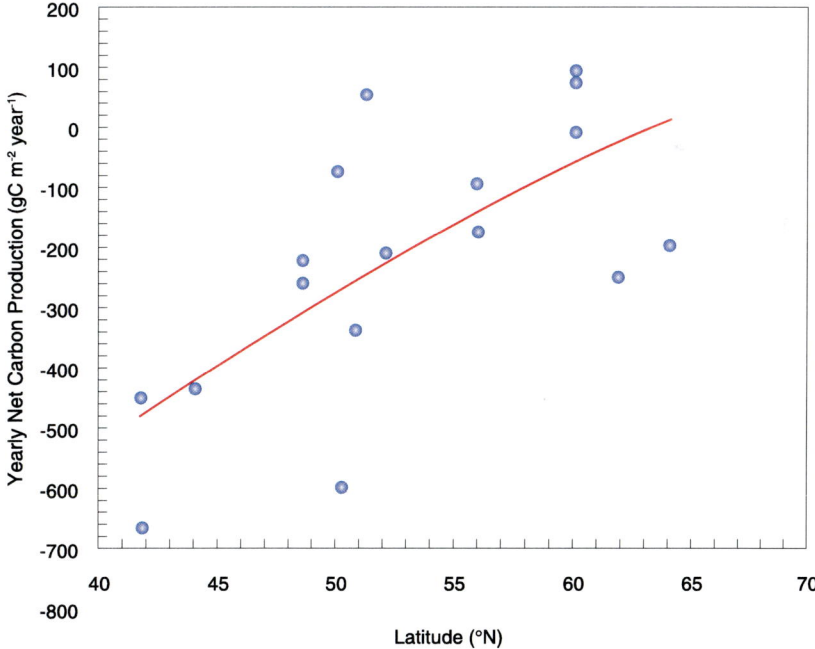

Figure 19. Net ecosystem carbon exchanges as measured by Euroflux (A minus figure represents absorption from atmosphere by forests). High latitude ecosystems can be a source of atmospheric carbon dioxide (Ronald, 1998).

Forest Sequestration Models and their Scaling-UP

We must establish a model for each of the processes shown in Fig. 17 before we can use monitoring data from the forests as a basis for understanding the distribution of CO_2 absorption amounts in the terrestrial ecosystem and to predict how it will be affected by possible future climate changes. A good example is Sim-CYCLE, a model developed in Japan. Assuming each box of leaves, stems and roots of plants, fallen leaves and branches, and inorganic soil, the amount of photosynthesis and plant respiration, the volume of fallen leaves and branches, and the rate of decomposition in the soil can be calculated by inputting environmental factors such as the CO_2 concentration, temperature, and humidity as parameters for each type of vegetation. This long-term data will eventually reveal the naturally stable capacity for carbon storage. Figure 20 illustrates the calculation results for Asia. Figure 21 reveals the estimated increase of net primary productions (NPP, see Fig. 17) when CO_2 emissions are doubled over the next 70 years.

However, a reliable model for future predictions has yet to be developed since the actual CO_2 sequestration amount will depend on various factors that are interconnected in complex ways, including climate changes (temperature rise, rainfall change, etc.) caused by a CO_2 increase as well as consequent vegetation changes.

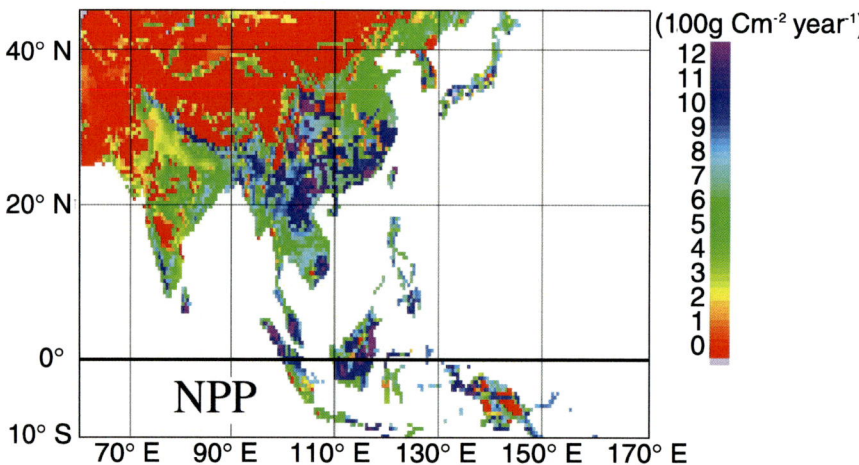

Figure 20. Regional distribution of NPP in Monsoon Asia estimated by equilibrium run of Sim-CYCLE Model (Oikawa and Ito, 2001).

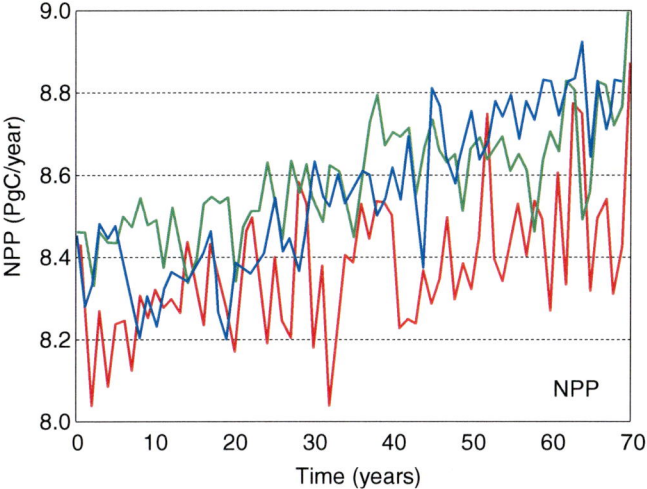

Figure 21. Projected time trends of NPP in Monsoon Asia using the tree GCM scenarios based on doubling of CO_2 concentration during a 70-years period (1%/year). Red :GFDL, green :GISS and blue :MRI (Oikawa and Ito, 2001).

Confirming the Reliability of Carbon-cycle Models – The Atmospheric Monitoring Approach

Individual tower observations enable us to evaluate the carbon cycle for a specific area and plant species. The results of these observations cannot be applied to other areas on different latitudes or those subject to different climatic conditions, or to different plant species. The measurements of short-term stock change are not sufficiently accurate. It is important to compile monitoring data from diverse plant species from different climates and latitudes and attempt a comprehensive interpretation to estimate the global amount of carbon sequestration by the terrestrial plant ecosystem.

We must therefore consider establishing long-term observation and research networks for the amount of CO_2 exchanges (flux) between the atmosphere and forest ecosystems positioned at different latitudes and in different environments. Surveys on climatic conditions, forest ecology (including vegetation), photosynthetic reactions and soil decomposition, in addition to satellite images, provide vital information and must be combined with flux observations.

Preparations are underway to establish a worldwide network for interdisciplinary cooperation by scientists (Fluxnet) to promote these research activities. Forests in East Asia exist under more complex conditions than those in Europe and US; these conditions include monsoons, high humidity, and heavy rainfall in the growth period. Thus, a flux-monitoring network in Asia (AsiaFlux) is being organized by cooperation among related institutions and universities in Japan, Korea, and China. (http://www-cger2.nies.go.jp/asiaflux/index.html)

In addition to these initiatives, attempts are underway to estimate where and how much CO_2 is emitted and absorbed, based on the data from the observation network for atmospheric CO_2 concentrations. CO_2 emission and absorption take place on the surface; the CO_2 in the atmosphere then is transported and gets mixed as the air flows, which determines the global CO_2 distribution. The CO_2 concentration will be greater upwind than downwind if the forest functions as a CO_2 sink, and therefore the amount absorbed can be estimated by a calculation using the wind velocity and the concentration difference.

Figure 22 depicts the estimated CO_2 balance in sub-continental regions, evaluated based on the CO_2 concentration figures supplied by the worldwide CO_2 monitoring network (the inverse model). Each number indicates CO_2 emissions in that area (a minus figure represents CO_2 absorption). The red numbers are estimates that incorporate atmospheric monitoring data for Siberia gathered by airplanes, which will be described later.

Reliable values of absolute concentrations are vital for analyzing this data since the absorption amount is estimated based on the minimal variances between different observatories. Relevant monitoring institutions are presently organizing a unified standard gas system to ensure accuracy in their observed values. Scientists are also considering establishing similar high-level observation networks for methane, isotopes, and oxygen, in addition to CO_2.

Another important point is that the current monitoring network focuses primarily on developed countries and many areas remain unexamined, such as tropical Africa, Southeast Asia, the northern part of South America, and the Eurasian continent. Airplanes are a valuable means to monitor the midair atmosphere to represent the concentration in the continent, since the surface concentration is directly influenced by nearby forests and fossil fuel consumption.

The red figures in Fig. 22 illustrate the results when aircraft data obtained over Siberia by the National Institute for Environmental Studies is also examined. This new data indicates that upwind areas in Europe and Siberia have higher levels of absorption, which effectively reduced the margin for error in the estimates. It is therefore important to fill the gaps in the observation network.

Analysis is presently only possible on an approximate basis, since resolution is only available at a scale of several thousand kilometers. Therefore, it is necessary to estimate the CO_2 absorption for each country separately by using data from closely connected air monitoring networks and comparing the results to the estimates provided from models based on ground monitoring data, such as flux observations. Observations over wide areas using advanced technology, such as satellites, are required. Satellite observation data would be very useful for high spatial resolution analyses on a 1000km scale, even though the estimated accuracy of a column concentration is 1% or better and no height distribution can be resolved. Ministry of the Environment and Japan Aerospace Exploration Agency (JAXA), Japan, plans to launch the Greenhouse Gas Observation Satellite (GOSAT) in 2007 for column concentration measurements using the differential absorption method.

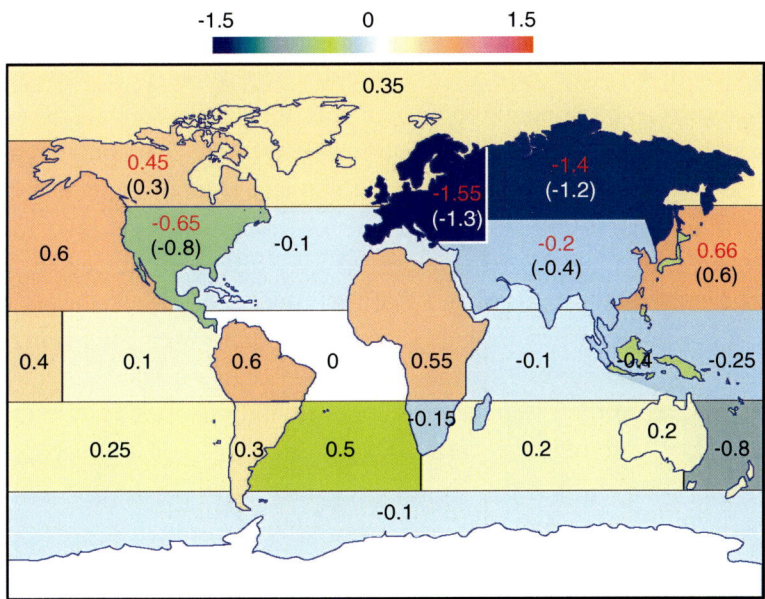

Figure 22. Annual carbon budget distribution estimated by Inverse Model analysis of global carbon dioxide observation database combined with atmospheric transportation database. Values are annual flux (PgC year^{-1}) in the regions (Rayner *et al.*, 1999) Red values are the new values when the aircraft observation data over Surugut (Siberia, by National Institute for Environmental Studies) was added (Maksyutov *et al.*, 2003).

1.4.3 The Ocean: A Large Sink for Carbon Dioxide Accumulated in the Atmosphere

The oceans absorbed an estimated 1.7PgC of CO_2, out of a total 6.3PgC emitted annually from burning fossil fuels, during 1990 and 2000. This is about half of the total amount of CO_2 of anthropogenic origin after subtracting the amount accumulated in the air (3.3PgC). Therefore, the oceans act as a large CO_2 sink, together with the terrestrial ecosystem.

Why do the oceans absorb CO_2?

CO_2 transforms into four different chemical forms when it is dissolved in water, carbonic acid, bicarbonate ions, carbonate ions, and a fraction remains as CO_2. The proportions of these chemical forms are determined by the pH of the water.

The pH of seawater is such that most CO_2 takes the form of bicarbonate ions (85%); the amount of total inorganic carbon dissolved can reach a level as high as 100 times greater than when it is in the form of CO_2. The oceans, with an average depth of 3800m, currently contain 50 times as much total inorganic carbon as atmospheric CO_2.

The CO_2 absorption rate is also accelerated by activities of marine organisms, such as phytoplankton near the ocean surface. Phytoplankton use inorganic carbonate to produce organic matter through their photosynthetic reaction. The organic matter produced in the upper layer of the ocean is then transported to the middle and deeper layers; biological decomposition of organic matter subsequently reproduces carbon dioxide in the deeper layers. These processes enhance the capacity of the middle and deeper layers of the ocean to store more carbon dioxide than the surface water. This function is known as "biological pump", storing CO_2 in the ocean interior.

Surface water of the ocean can both absorb and release CO_2 to and from the atmosphere. The balance between these processes of CO_2 absorption and release is determined by the difference in partial pressure between atmospheric and oceanic CO_2 (P_{CO_2}).

Atmospheric CO_2 is currently 370 µatom (CO_2 partial pressure, P_{CO_2}), while there is considerable seasonal variation in P_{CO_2} in ocean surface waters (150 to 750 µatom). This variation must be considered in all observations.

The standard device used to measure CO_2 in ocean surface water is an infrared gas analyzer equipped with an air-seawater equilibrator system to measure both atmospheric and oceanic CO_2 at the same time.

This monitoring system requires a ship equipped with the measuring instrument to cruise the area of interest, which makes it difficult to obtain concentrated data for the vast expanse of the oceans.

About one million samples of P_{CO_2} collected from surface water of the world's oceans have been analyzed since commencement of a project connected with the International Geophysical Year (1957-1958). The data from those measurements provided the basis for the P_{CO_2} Map, which illustrates the CO_2 flux between surface water and the atmosphere over the oceans in the world.

Figure 23 depicts the annual CO_2 absorption-release rate (flux) of global oceans, using the P_{CO_2} data and the CO_2 dissolution rate constant evaluated for each location. The positive flux rates at middle and sub-arctic latitudes indicate that the CO_2 partial pressure in surface water in these areas is lower than that in the atmosphere.

Ocean surface water cools as it flows northward from lower latitudes. This cooling process increases the capacity of surface water to dissolve CO_2 into the water, although the CO_2 absorption process takes time. This is the reason the CO_2 partial pressure is lower in middle and sub-arctic latitudes. A similar process occurs when phytoplankton sequesters CO_2.

In contrast, CO_2 fluxes from the ocean surface in the equatorial zone are high. Upwelling in this region brings seawater containing large amounts of total inorganic carbon from the ocean depths to the surface, which makes the CO_2 partial pressure greater than in the atmosphere. The results presented in Fig. 23 represent the data gathered over the past 40 years and provide the mean figures for P_{CO_2}.

There has been an increasing need for P_{CO_2} monitoring in recent years, corresponding to the potentially broad spatial and seasonal variations of CO_2 flux in the ocean. One approach has been installation of CO_2 measuring equipment on commercial vessels that cruise a regular route, which enables more intensive monitoring programs.

A good example of such surveys in Japan is the study of P_{CO_2} of sub-arctic Pacific surface waters conducted by NIES from 1995 to 2001 (Fig. 24).

The sub-arctic Pacific undergoes substantial seasonal variations, and P_{CO_2} varies widely in each season. The figures indicate that this area is a CO_2 sink from spring to summer, but becomes a CO_2 source in winter, with the Bering Sea at the center.

Observation ships of the Japan Meteorology Agency have conducted a similar monitoring program. The Agency measured P_{CO_2} along the transect at 137°E longitude. The seasonal changes are directly opposite those of the sub-arctic zone; the sub-tropics become a CO_2 sink from winter to spring.

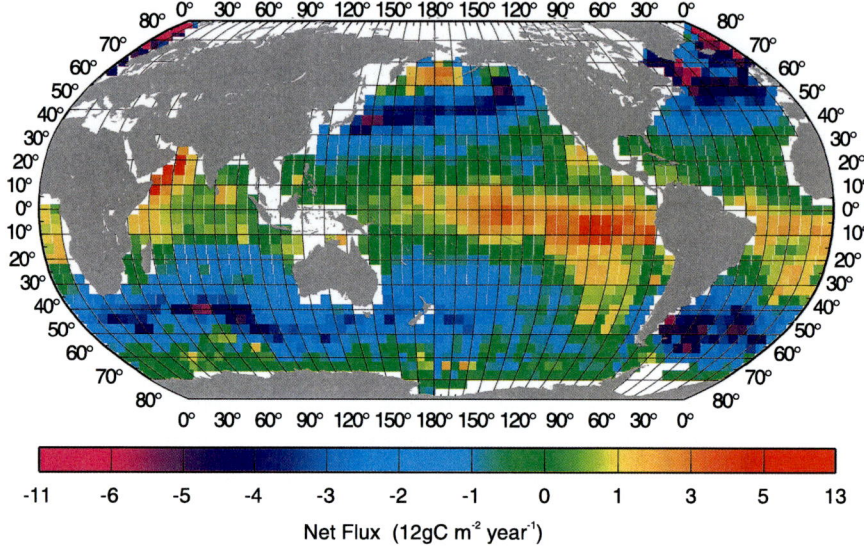

Figure 23. Distribution of the climatological mean annual air-sea CO_2 flux over the global ocean for the reference year 1995 (Feely *et al.*, 2001).

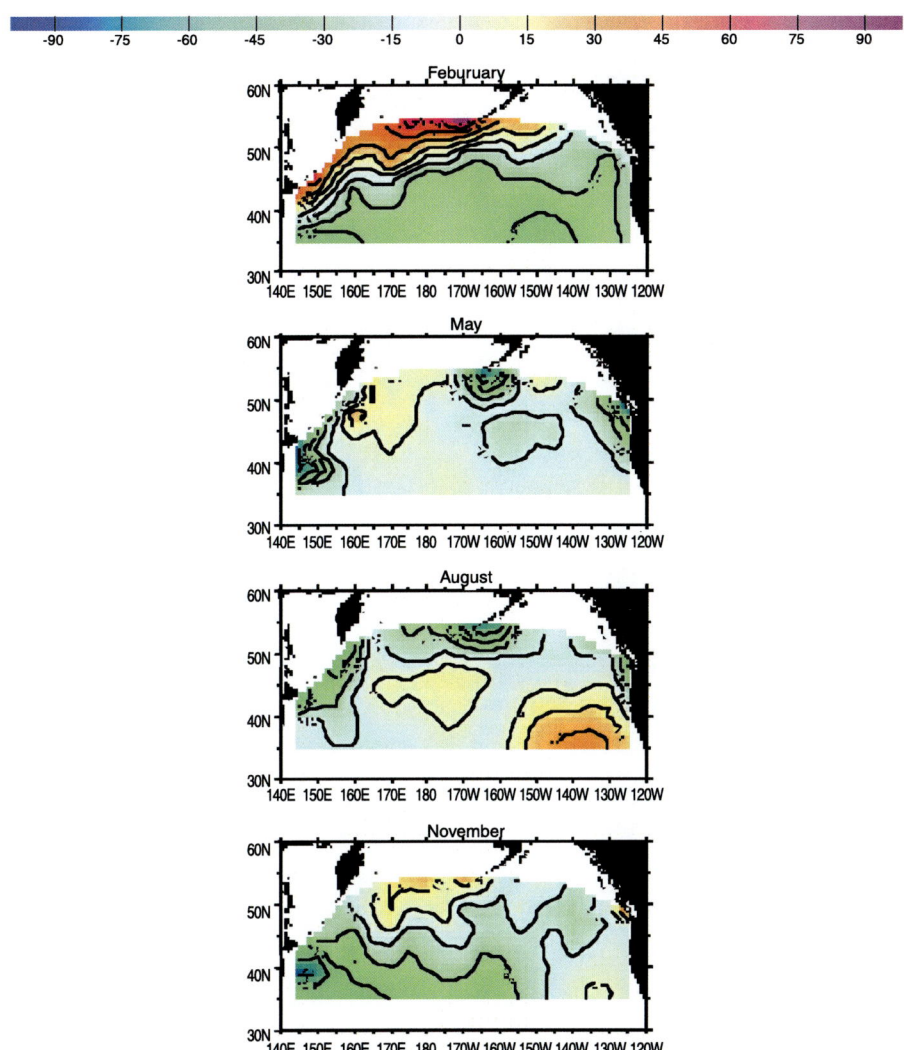

ΔP$_{CO_2}$ in the subarctic Pacific obtained by a joint project of the National Institute for Environmental Studies and Institute of Ocean Sciences, Canada using a cargo ship "Skaugran" from 1995 to 1999. Unit : μ atom

Figure 24. Seasonal variation in air-sea P$_{CO_2}$ distribution in the subarctic Pacific (Zeng *et al.*, 2002).

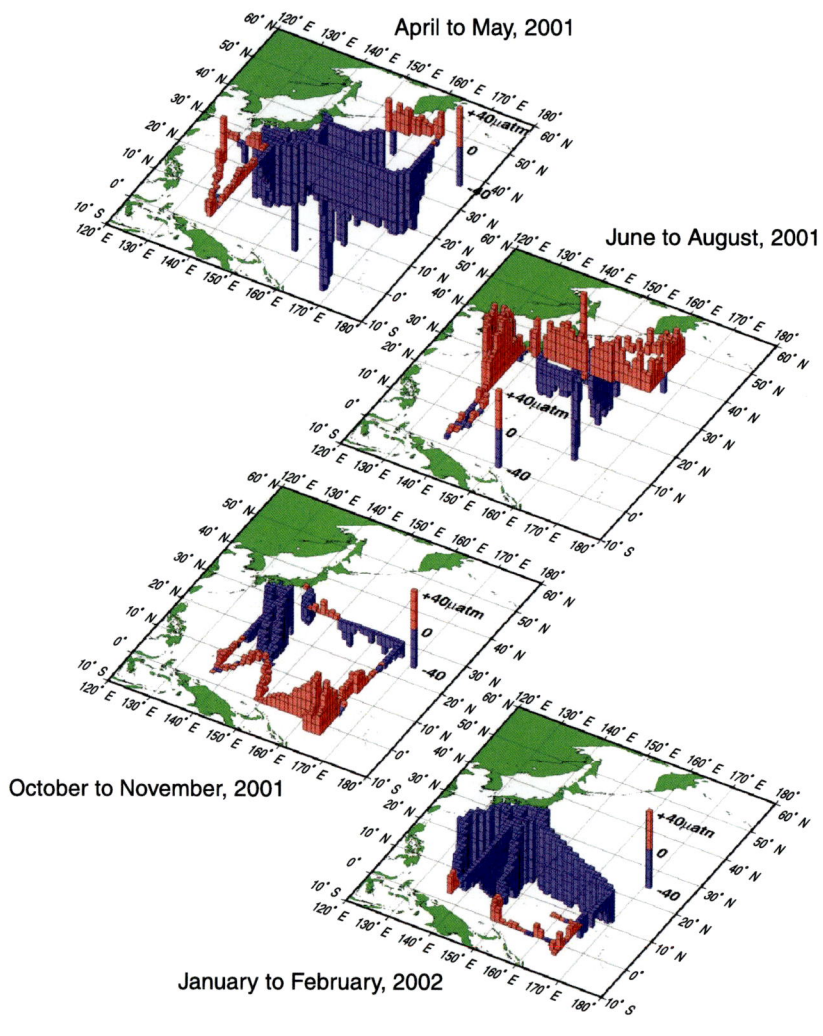

Figure 25. Spatial distribution of P_{CO_2} in the Western Pacific (Japan Meteorological Agency, 2003).

1.4.4 Why it is Important to Monitor the Ocean Carbon Cycle

P_{CO_2} in seawater is chemically determined by four factors: seawater temperature, salinity, pH, and the concentration of total inorganic carbon.

These factors are affected by physical factors such as heat exchange, freshwater exchange (evaporation, rainfall, etc.), and changes in surface seawater circulation due to wind stress. All these factors are related to P_{CO_2} changes in the ocean surface.

For example, CO_2 is more easily dissolved in winter when temperatures are low in middle-latitude oceanic areas, where water temperatures exhibit wide variations, since dissolution of gas components like CO_2 into seawater decreases according to increases in water temperature.

The concentration of total inorganic carbon and the pH of seawater are heavily influenced by biological activities, such as CO_2 fixation by phytoplankton, generation of inorganic carbon by biodegradation of organic matter, and formation of calcium carbonate, such as shells of foraminifera. The surface-produced organic matter and the calcium carbonate shells sink to the middle or deeper layers of the ocean, contributing to CO_2 absorption from the atmosphere at surface layers.

Biological activities in ocean surface water thus significantly impact the absorption and release of atmospheric CO_2, similar to terrestrial plants. Parameters for the carbon cycle related to CO_2 increase/decrease in surface waters are therefore important for studies of global warming, which is why many studies in recent years have focused on carbon cycles.

The primary subjects of recent studies include biological parameters, such as the amount of chlorophyll in phytoplankton and the concentration of nitrate as an important nutrient. These parameters are observed using a satellite directly for chlorophyll or indirectly for nutrients through the correlation with seawater temperature.

Figure 26 depicts the results of monitoring chlorophyll concentrations in global surface water using satellites equipped with an ocean color sensor. The satellite monitoring in this figure was first conducted by a Japanese satellite sensor (NASDA OCTS) in 1997, which was succeeded by a US satellite (NASA SeaWiFS) in 1998. The concentration becomes greater as the color changes from blue to red. 1998 was an El Niño year, and 1999 was a La Niña year. A clear difference in chlorophyll concentration can be seen around the eastern equator between the two observations, even though they are both for the same month, January.

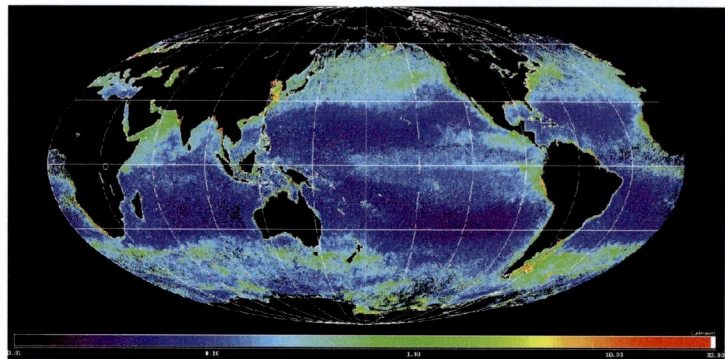
January, 1997

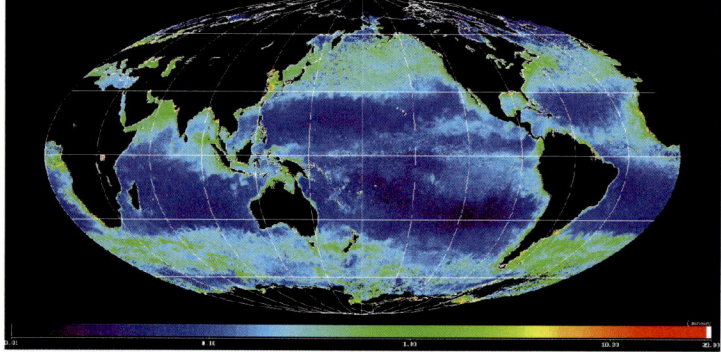
January, 1998

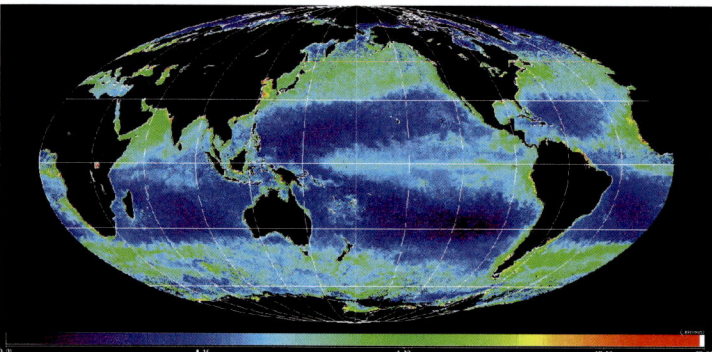
January, 1999

Figure 26. Annual variation of surface chlorophyll concentrations over the global ocean evaluated from the satellites with ocean color sensors of NASDA (currently known as JAXA) and NASA.

Scientific Survey Ships and Fixed-Point Observations of the Carbon Cycle

Time-series observations of the marine carbon cycle at fixed locations are currently conducted most intensively at two monitoring sites, offshore Hawaii in the Pacific (HOT) and offshore Bermuda in the Atlantic (BATS). The Joint Global Ocean Flux Study (JGOFS), an international project undertaken to study marine biogeochemistry as related to global environmental changes, manages these two programs, which have recorded detailed seasonal changes of carbon cycles for more than ten years.

Figure 27 depicts the long-term changes in total inorganic carbon at the surface water in these two locations. Increases of 1.18m mol kg^{-1} per year at HOT and 1.23m mol kg^{-1} per year at BATS were observed in response to an increase in atmospheric CO_2.

Japan has also had a fixed monitoring station since 1999, KNOT, located offshore in eastern Honshu. However, it is very difficult to run a monitoring program in Japan, since there are no ocean islands sufficiently equipped with the necessary facilities and with easy access to open ocean, as found in Hawaii. KNOT is located well offshore and beyond reach of any impact from the coast, but it takes many days for researchers to travel to the site and back.

There are presently about ten fixed monitoring sites in global ocean areas, including observations of carbon cycle. However, they differ significantly in regard to what they monitor and how often they take samples.

Most fixed stations perform time-series observations in the oceans by collecting their data using unmanned buoy systems. A typical example is the buoy system for monitoring El Niño at the Pacific equator, which contributes significantly to forecasting the El Niño event, a phenomenon closely related to global climate changes. The US is responsible for the eastern part of the Pacific, and a Japanese marine research ship, "Mirai" ("future" in Japanese) covers the western part.

These buoys primarily monitor physical parameters such as water temperature, salinity, and water flow. Long-term monitoring of chemical elements is also crucial for time-series observations of the carbon cycle, but unmanned sensors capable of this task have yet to be fully developed. This is a challenge that we presently face.

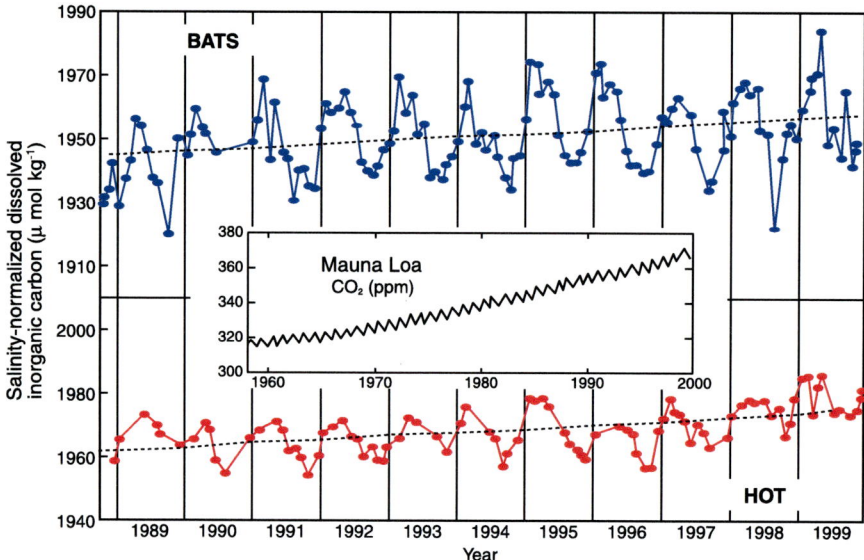

Figure 27. The 11-year record of mixed-layer dissolved inorganic carbon (DIC), normalized to a salinity of 35.0 for the two US-JGOFS time series stations (BATS and HOT) (Karl *et al.*, 2001).

CHAPTER 2

CLIMATE MODELING AND THE PROJECTION OF GLOBAL WARMING

2.1 History of the Projection of Global Warming

Unlike numerous other environment or pollution issues, the global warming issue has spurred society to action without any tangible "issue" at hand, based upon future projections and warnings provided by scientists. It is therefore essential for the decision-makers of society to learn the scientific basis of the global warming projection. The "science of global warming" could be divided into two areas: research on the anthropogenic increase of CO_2 concentration and research on climate change due to such an increase. In this chapter, we will concentrate on the latter where future projection by supercomputer simulations plays a major role.

The Earth warms up by absorbing solar radiation energy on the one hand and by emitting infrared radiation energy according to its surface temperature on the other, thereby maintaining its energy balance. If we consider only these processes in calculating the equilibrium temperature of the Earth's surface, we get -18°C as the answer. However, the actual average surface temperature of the Earth is +15°C, 33°C higher than the value calculated by radiation balance.

The difference is due to the following reasons. The Earth's atmosphere allows the incoming solar radiation to pass through it with little absorption, while some of the atmospheric constituents absorb outgoing infrared radiation from the Earth's surface and then re-emit further outgoing radiation in addition to downward radiation which also warms up the Earth's surface. The Earth's surface, thus additionally warmed, emits more infrared radiation, which, however, is absorbed by the atmosphere. In the eventual total balance, the Earth system is seen from space to emit infrared radiation from the surrounding atmosphere with the same temperature of -18°C if the absorbed solar radiation is not changed.

In the early 19th century, Jean B. J. Fourier [*1], a famous mathematician, argued for the first time the possibility of calculating the surface temperature of the Earth by using the radiation balance as described above. J. Tyndall

[*2], a British physicist, then first realized that water vapor and carbon dioxide are atmospheric constituents that absorb infrared radiation from the Earth. From the last years of the 19th century to the early years of the 20th century, S. A. Arrhenius [*3], a Swedish physical chemist, argued the possibility that the change in the atmospheric concentration of carbon dioxide could have brought about a large-scale change in the surface temperature of the Earth (i.e., glacial and interglacial periods) and also inferred that the anthropogenic emission of carbon dioxide would cause global warming. Later studies were made to qualitatively evaluate the change in the surface temperature of the Earth due to the increase of the atmospheric concentration of carbon dioxide, but they considered only the radiation energy balance.

Both the effect of radiation and that of convection were then considered by Manabe and Wetherald (1967), who successfully established a firm basis for evaluating global warming due to the increase of the atmospheric concentration of carbon dioxide. Their one-dimensional radiative-convective equilibrium model has been developed into today's coupled atmosphere-ocean models (three-dimensional climate models), which are now fully utilized to project global warming.

Figure 1 depicts the projected global warming in the Third Assessment Report (TAR) of the Intergovernmental Panel on Climate Change (IPCC, 2001a) based upon recent research results. The globally averaged surface temperature is projected to increase by 1.4 to 5.8°C from 1990 to 2100. The global mean sea level is also projected to rise by 0.09 to 0.88 meters over the same period.

In this chapter, we first review how such global warming has been projected and to what extent the regional features of global warming have been assessed as the globally averaged surface temperature increases. Second, we will present major international research programs under which global warming research is being conducted worldwide. Finally, we summarize the history and future directions of Japanese climate modeling studies of global warming.

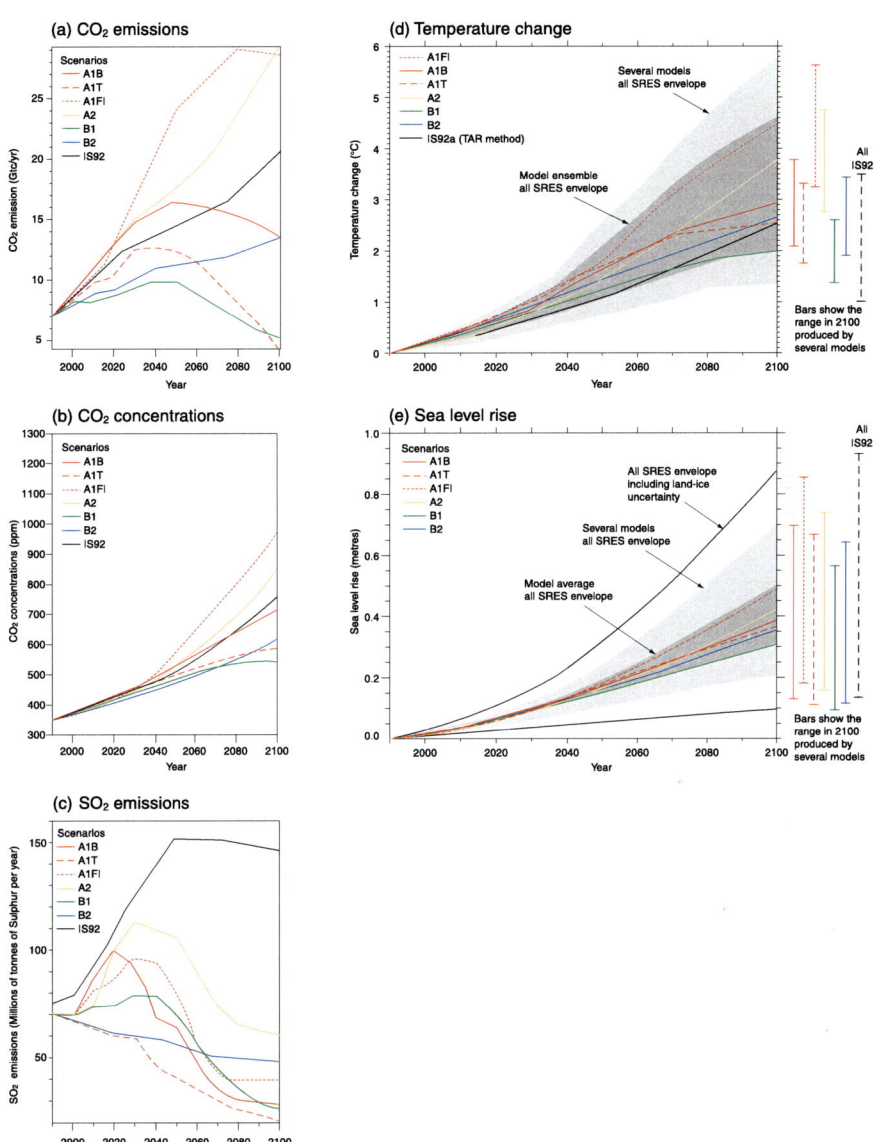

Figure 1. Six illustrative SRES scenarios for (a) CO_2 emissions, (b) CO_2 concentrations and (c) SO_2 emissions used in IPCC TAR, and (d) temperature and (e) sea level rise responses projected by the simple model tuned to a number of complex models with a range of climate sensitivities (IPCC SPM, 2001a).

2.2 How is Global Warming Actually Projected?

2.2.1 Earth's Climate System

What characterizes the climate of the Earth on which we live? The TAR of the IPCC defines the climate system as follows.

"The climate system is the highly complex system consisting of five major components: the *atmosphere*, the *hydrosphere*, the *cryosphere*, the *land surface*, and the *biosphere*, and the interactions between them. The climate system evolves in time under the influence of its own internal dynamics and because of external forcings such as the changing composition of the atmosphere and *land-use change*."

The above definition emphasizes the role of the interactions between the components of the climate system. However, every system has a specific function. What, then, is the most important function of the Earth's climate system?

More solar energy is absorbed on the Earth's surface than in the atmosphere in the vertical direction since the Earth's atmosphere is highly transparent to solar radiation. More solar energy is also absorbed in equatorial regions than in polar regions horizontally since the Earth is a sphere. The energy balance, however, cannot be maintained by radiative cooling alone (i.e., emission of infrared radiation energy) in areas that absorb excess solar energy. There, the climate system of the Earth converts extra solar energy into heat energy and transports it to regions that are absorbing relatively less solar energy, where infrared radiation energy is emitted into space to maintain the overall energy balance of the system. Thus the main function of the Earth's climate system is to transport heat to maintain the heat balance [see Fig. 2].

The vertical heat transport creates a layered structure called the troposphere, while the horizontal transport forms organized, large-scale circulations in the atmosphere and ocean, such as the trade winds in the low latitudes and westerly jets in the middle-high latitudes, monsoon circulations, and the Gulf Stream. These circulations are affected by the Earth's rotation and the land-sea contrast. The heat transport necessarily accompanies momentum and material transports.

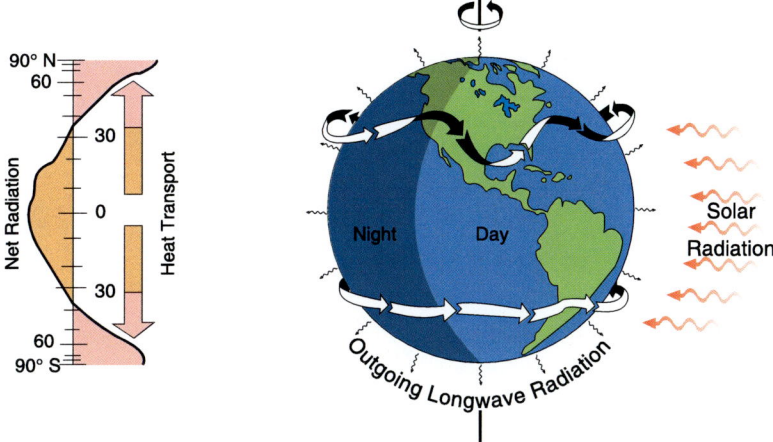

Figure 2. Energy balance and poleward heat transport of the Earth (IPCC, 1996).

2.2.2 Climate Model

A climate model is a numerical representation of the climate system used to calculate properties of its components, their interactions and feedback processes, and their temporal evolutions.

First, the model covers the atmosphere, ocean and land surface with a regularly distributed grid point system. It then calculates the budgets of energy, momentum and mass among these grid points. In addition fluxes with scales smaller than the grid interval also have to be estimated since they are not negligible for the budget calculations.

However, there is no established way of representing the above effects in terms of grid scale physical quantities (such representation is referred to as "parameterization" in meteorology). Hence, various parameterizations have been devised and applied for each physical process in existing climate models. Because of the difference in parameterizations, there are a number of different climate models, producing different results for a set of prescribed experiment conditions. In particular, the difference in the parameterization of cloud-related physical processes causes the largest scatters in projected results of global warming among existing climate models. Improving such parameterization is now the highest priority for further developing climate models.

2.2.3 The Earth will not Warm Homogeneously

The carbon dioxide emitted by various anthropogenic activities is geographically distributed. However, the atmospheric concentration of carbon dioxide increases almost homogeneously worldwide because its lifetime is long enough for it to be mixed by atmospheric general circulations. Nevertheless, the Earth is not projected to warm homogeneously. Figure 3 illustrates an example of the distribution pattern of such global warming projected by the climate model developed at the Meteorological Research Institute (MRI). Though there are some differences in the projected results of global warming among existing climate models, they all exhibit similar features in the distribution pattern of global warming. The past three assessment reports of the IPCC confirm the distribution patterns depicted in Fig. 3 as qualitatively common among climate models. Some mechanisms which cause such characteristic features have already been clarified by analyzing the model results, providing a basis for further study in detecting and attributing global warming.

However, several components of the climate system (e.g., interaction between aerosols [*4] and clouds, land surface vegetation, sea ice, carbon cycle, etc.) have not yet been sufficiently modeled. Furthermore, some climate processes (e.g., dynamics of continental ice sheets, etc.) have not been incorporated in any climate models. Thus, there may still be differences between the actual climate change in the future and the global warming simulated by existing climate models. Therefore, we need to continue improving climate models and verifying them by observations.

Characteristics of Global Warming Projected by Climate Models
Major characteristics of global warming projected by existing climate models are summarized below.

Projected changes in the vertical direction:

- In the troposphere, the temperature increases due to the additional anthropogenic greenhouse effect, while it decreases in the stratosphere to balance the heating due to the absorption (by O_3) of the solar radiation and the cooling due to the emission (by CO_2, O_3, etc.) of infrared radiation. The recent decrease of Ozone may also contribute to

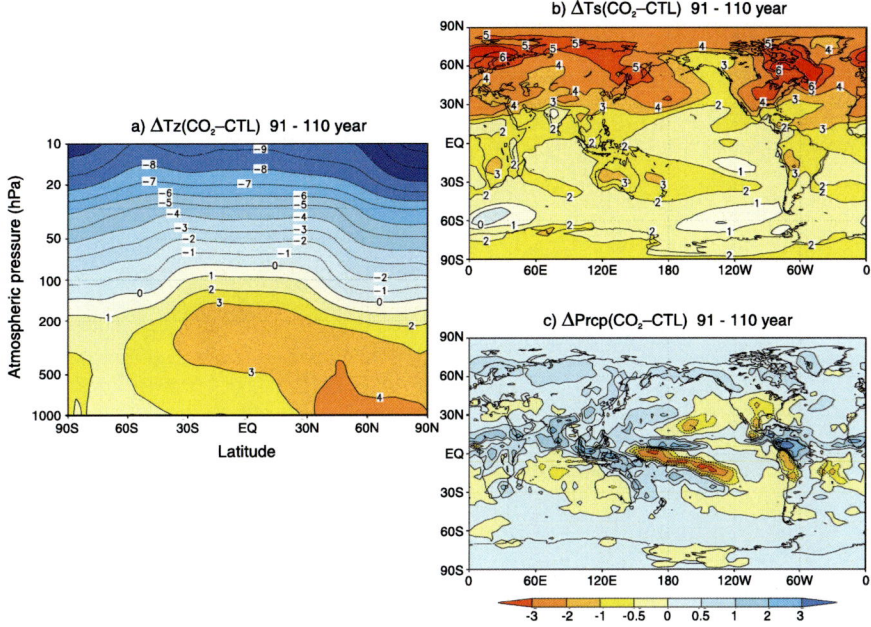

Figure 3. Changes in a) meridional distribution of temperature (in degrees Celsius), b) geographical distribution of surface air temperature (in degrees Celsius) and c) geographical distribution of precipitation (in mm/day), for 1%/yr CO_2 increase around the time of CO_2-tripling (20-year mean over years 91 to 110), simulated with the MRI AOGCM (Noda, 2000).

the decrease in temperature in the stratosphere.
- In the arctic area, the heat flux from the ocean to the atmosphere will increase due to the reduction of sea-ice extent and thickness in winter, the absorption of solar radiation will increase due to the snow- and ice-albedo feedback effect in summer. These effects contribute to the larger increase in temperature in the lower troposphere over the arctic area. Since the atmosphere stratifies very stably in winter, the warming does not affect the upper layers.
- In low latitudes where cumulus convection is active, the temperature lapse rate is close to the moist adiabatic lapse rate [*5]. This lapse rate decreases as temperature increases. Therefore, global warming causes a larger temperature increase in upper layers than in lower layers.

Projected changes in the horizontal direction:

- There will be a significant asymmetry in the increase of surface temperature in high latitudes between the Northern and the Southern Hemispheres. A larger increase will occur in high latitudes of the Northern Hemisphere due to the sea ice and snow feedback effects as noted above. However, the surface ocean current sinks to deep layers near the Antarctic Continent making the feedback of sea ice and snow ice less effective, so the sea-surface temperature will not increase much even if heat is input near the sea surface.
- Since the evaporation increases rapidly as the temperature increases, small increases in sea surface temperature can adjust the enhanced downward infrared radiation at the surface by enhanced evaporation (upward latent heat flux). In addition, the heat capacity is larger for sea water than for land soil. These two effects result in a larger temperature increase over land than over the ocean in the same latitudinal zone.
- The poleward transport of water vapor will increase so that precipitation will increase in high latitudes.
- The low-level easterly circulation prevailing in the equatorial Pacific normally becomes upward motion at its western boundary where cumulus convective activities are dominant due to higher sea-surface temperature. It then changes to westerly circulation heading to the east in the upper troposphere, and finally becomes downward motion in the eastern equatorial Pacific where the sea-surface temperature is lower, forming a closed circulation system called the Walker Circulation. This circulation is weakened during an El Niño event and strengthened during a La Niña event. If global warming strengthens the circulation, similar changes in sea-surface temperature, precipitation and pressure would prevail as in a La Niña event, whereas the opposite changes as in an El Niño event would occur if global warming weakens the circulation. In the Meteorological Research Institute (MRI) climate model, the projected pattern of global warming exhibits changes similar to a La Niña case, while other climate models project changes similar to an El Niño case.

2.3 Worldwide Research Efforts on Projecting Global Warming

The hot summer of 1988 in the USA dramatically increased the momentum for turning the global-warming issue into one of the most important issues in the international political arena, leading to the establishment of the Intergovernmental Panel on Climate Change (IPCC), an international body for assessing scientific, technical and socio-economic information relevant for understanding climate change, its potential impacts, and options for adaptation and mitigation.

2.3.1 GCM Experiments for IPCC

To project global warming, the emission scenarios of greenhouse gases (e.g., CO_2) and aerosols are incorporated into a climate model for computing future projection (such a computing process using a numerical model is generally called a numerical experiment). Thus computed results are analyzed further for assessing the impact of global warming. The progress of these efforts and activities is summarized in Table 1.

In the past 10 years, climate modeling, which started from atmospheric general circulation models (AGCMs) coupled with a thermodynamic slab ocean, has progressed to developing coupled atmosphere-ocean general circulation models (AOGCMs). Based on developments of the methodology, the earlier "doubled CO_2 equilibrium experiments," where the atmospheric concentration of CO_2 is fixed to twice the standard concentration, have been replaced by "transient experiments" where transient responses are investigated by gradually increasing the atmospheric concentration of CO_2.

Table 1. Approximate chronology of IPCC process in relation to GCM simulations, their adoption in impact studies, and the development of IPCC emissions scenarios. Abbreviations: AGCM atmospheric GCM with simple ocean; AOGCM coupled atmosphere-ocean GCM; GHG greenhouse gas; IS92 IPCC emissions scenarios published in 1992 (Leggett *et al.*, 1992); SRES Special Report on Emissions Scenarios (Nakicenovic *et al.*, 2000) (IPCC 2001b, Table 3-6).

Date	IPCC Process	Working Group I GCM Simulations	Working Group II GCM-Based Scenarios used in Impact Studies	Working Group III Emissions Scenarios
1988-1990	First Assessment Report (FAR), 1990	Equilibrium high-resolution AGCM	Equilibrium low-resolution $2 \times CO_2$	Scenarios A-D (A = Business-as-Usual)
1991-1992	FAR Supplement, 1992	Transient AOGCM cold start GHG-only (Scenario A emissions)	Equilibrium low-resolution $2 \times CO_2$	IS92a-f
1993-1996	Second Assessment Report (SAR), 1996	Transient AOGCM warm-start GHG + aerosol (0.5 or 1% per year emissions)	Equilibrium low/high-resolution; transient cold-start	IS92a-f (modified)
1997-1998	Regional Impacts Special Report, 1998	Transient AOGCM ensemble/multi-century control	Equilibrium low/high-resolution; transient cold-start/warm-start	IS92a-f (modified)
1999-2001	Third Assessment Report (TAR), 2001	Transient AOGCM CO_2-stabilization; SRES-forced	Transient warm-start; multi-century control and ensembles	SRES; stabilization

2.3.2 Development of Climate Modeling

Climate modeling efforts are being pursued in each climate research center. International cooperation among the research centers is organized by the World Climate Research Programme (WCRP).

The WCRP consists of the following four subprograms:
- Global Energy and Water Cycle Experiment (GEWEX),
- Climate Variability and Predictability Research Programme (CLIVAR),
- Climate and Cryosphere of Arctic Climate System Study (CliC) to follow up Arctic Climate System Study (ACSYS), and
- Stratospheric Processes and their Role in Climate (SPARC).

Each of these has a working group on numerical modeling. The one closely related to climate modeling is the Working Group on Climate Model (WGCM) in CLIVAR; it focuses on coupled modeling activities.

> **Model Intercomparison**
>
> Various international research projects are now ongoing to promote climate modeling. Participating groups conduct numerical integration experiments under the same experimental design and then compare the properties among their models and/or study the mechanism of climate variability and change based on a comparison of their experimental results. Such projects usually have the words Model Intercomparison Program (MIP) in common in their titles.
>
> One of them, the Coupled Model Intercomparison Project (CMIP), reviewed and fostered by the WGCM, has been playing a central role in modeling for predicting climate change. The IPCC is focusing on the scenario-based projection of global warming as a necessary basis for policy making, while the CMIP is trying to clarify the mechanism of global warming scientifically through such experiments as doubling CO_2 experiments or gradual (1% per year) CO_2 increase experiments.

In the forthcoming Fourth Assessment Report of the IPCC, the application of higher resolution will be a major focal issue for climate modeling along with the inclusion of carbon cycles. In the early stage of climate modeling, the atmospheric resolution was a horizontal grid interval of about 500km. In the Second and Third Assessment Reports of the IPCC, an atmospheric resolution of about 250km was applied to each climate model. Since the regional detailed aspects of global warming and its impacts on extreme weather events have raised increasing concern as issues to be assessed, efforts are being made for climate modeling with finer resolution. For example, in subprojects under a national project "Sustainable Co-existence Project on Human, Nature and the Earth" (details to be presented later), research is now proceeding on projecting global warming by high-resolution climate models.

In the United States, a unified modeling initiative, the Community Climate System Model (CCSM), has recently been undertaken by integrating two major climate models (http://www.ccsm.ucar.edu/). Furthermore, a pilot project, "Programme for Integrated Earth System Modelling (PRISM)" has recently been launched in the European Union to establish a climate research network infrastructure to promote Earth system modeling and climate prediction (http://prism.enes.org/).

2.4 National Efforts in Projection Research on Global Warming

2.4.1 History of the Projection of Global Warming in Japan

The computer simulation of climate change has been evolved from numerical weather prediction, which is now the main basis for daily weather forecasting. Numerical weather prediction was born in the USA with the electric computer developed in the 1950s and was soon introduced to Japan. Equipped with the most advanced computer at the time, the Japan Meteorological Agency (JMA) initiated numerical weather prediction operationally in 1959, the third organization in the world to do so.

However, in a real sense, numerical weather prediction became operationally usable worldwide in the 1980s, thanks to the development of computers enabling the computation of complex meteorological models and to the establishment of the global and around-the-clock monitoring system of meteorological satellites. Since the launch of the geostationary meteorological satellite, "Himawari" No. 1, in 1977, Japan (through the JMA) has been cooperating with the USA and Europe in meteorological satellite observation. JMA has been successful in numerical weather prediction, comparable with USA and Europe.

Research into climate simulation by extending the prediction period in numerical weather prediction to one month, one year and so on (and including the effect of changes in the ocean and on the land as well as in the atmosphere) was initiated in parallel with numerical prediction only in large research institutes in the USA because only they could afford to perform the huge amount of computations needed and also because such research was not considered very urgent in the basic research areas of Earth science.

In the 1970s, it gradually became the consensus among climate researchers that global warming was actually taking place, while society, for its part, became more sensitive to such natural fluctuations as extreme weather events because of the large scale expansion of human and industrial activities. Under these circumstances, concerns about global warming were raised in society and the government. Subsequently, climate simulations were conducted in many research institutes in the USA. The simulations were also started in Europe in several newly founded institutes in the 1970s.

In Japan research activities in universities were restricted to basic areas to understand mechanisms in atmospheric and/or oceanic phenomena using simplified models, partly because of insufficient computer resources. The Meteorological Research Institute (MRI) of the JMA was the sole institute in Japan where preparatory research activities were pursued for the possible substantial development of climate modeling in the future. With the relocation from Tokyo to Tsukuba in 1980, the MRI equipped a fully available computer for research purposes and climate modeling was finally launched.

The global warming issue popped up as an international political one with the 1988 heat wave in the USA. It also became widely known that the issue was based upon the outcomes of computer-based climate simulation. In the 1990s, institutions involved in the issue in Japan introduced computers and strengthened their efforts in climate research activities.

In 1990, the National Institute for Environmental Studies (NIES) introduced a supercomputer for cooperative use among institutions of all ministries or agencies. In 1991, the University of Tokyo established the Center for Climate System Research (CCSR) for cooperative research among universities. In 1997, the Frontier Research System for Global Change (FRSGC) was launched as a cooperative research project between the National Space Development Agency of Japan (NASDA) and the Japan Marine Science and Technology Center (JAMSTEC), both under the Science and Technology Agency (now merged into the Ministry of Education, Culture, Sports Science and Technology (MEXT)). At the same time, the "Earth Simulator" Project was also launched to create the most advanced parallel supercomputer in the world. The project was completed early in 2002, and the computer became operational in March 2002 (see Fig. 4).

The research community in Japan, having started its research activities 10 years behind the USA and Europe, is now trying to catch up with the world's advanced level.

Figure 4. Earth Simulator in operation.

2.4.2 Research on Global Warming Projection by Global Climate Models

In Japan, computer availability was limited as described above, so climate change experiments were conducted in the 1980s using climate models developed by MRI and CCSR/NIES. The Frontier Research System for Global Change Research initiated studies of the mechanisms of global warming with the participation of Dr. Shukuro Manabe, a leading scientist in global warming. Some results of the above research activities are presented below.

Meteorological Research Institute (MRI)

At its relocation from Tokyo to Tsukuba in 1980, the Meteorological Research Institute (MRI) installed a new super computer (with the fastest computing power in Japan at that time), which enabled the development of an atmospheric general circulation model (AGCM). A number of climate sensitivity experiments were performed using the model.

Since the late 1980s when global warming came to be of concern, a considerable portion of computing resources has been made available for experiments in global warming projection. Results from such experiments have been issued by the Japan Meteorological Agency (JMA) as Information on Global Warming Vols. I, II, III and IV, and their data have been digitized on CD-ROMs to utilize in global-warming impact assessment studies. The above results have also contributed to the past Assessment Reports of the IPCC as follows.

- Equilibrium doubled CO_2 experiments were conducted for the First Assessment Report of the IPCC (1990) by coupling an atmospheric general circulation model with a slab ocean model where only the effect of the heat capacity of the ocean about 50m in depth was incorporated [see Table 1]. The atmospheric part of the model included two types of precipitation, convective precipitation [*6] and large-scale precipitation [*7]. The effect of global warming on the distribution of precipitation frequencies and that of precipitation areas for each of the two types were analyzed. The results demonstrate that in a warmer climate the precipitation amount due to cumulus convection increases, but the global percentage of total areas of precipitation decreases. As a result, precipitation tends to concentrate. However, the result of an experiment conducted in Germany and reported in the Third Assessment Report of the IPCC indicates that the precipitation amount due to cumulus convection decreases in a warmer climate. These results suggest that the structural change in precipitation in a warmer climate may depend on the uncertainty in changes of atmospheric stability, large-scale atmospheric circulation and so on, leaving clarification as a future challenge.
- The outcomes of a gradual CO_2 increase experiment using a coupled atmosphere-ocean model, developed in the MRI as the first such model in Japan for the Second Assessment Report of the IPCC (1996), were reflected in the chapter titled "Climate Models – Projections of Future Climate." During the development of the MRI coupled atmosphere-ocean model, considerable efforts were concentrated on improving the model performance in simulating El Niño and Southern Oscillation (ENSO) events that are the most significant natural variability phenomena in Pacific regions. As a result, the developed model successfully simulated El Niño events and the decadal-scale

natural variability observed in Pacific regions. The spatial anomaly pattern of sea-surface temperature change (global warming pattern) assumes a pattern similar to the one that appears when a La Niña event takes place [left column of Fig. 5].
- In contrast, an equilibrium experiment using a coupled atmosphere-slab ocean model exhibits an ENSO-like natural variability. However, now the projected global warming pattern of sea-surface change assumes an El Niño-like spatial anomaly pattern, though the atmospheric part of the model is exactly the same as that of the above coupled atmosphere-ocean model [right column of Fig. 5]. The reason the two models exhibit opposite response patterns has not been clarified yet. However, such a similarity between global warming and natural variability patterns has been found in global warming experiments by a number of coupled atmosphere-ocean models worldwide and is referred to as an El Niño-like pattern of global warming.
- To contribute to the TAR, efforts were made to develop faster computing schemes to attain higher resolution in the climate model. Global warming experiments were conducted using greenhouse gas scenarios prepared by the IPCC for the TAR (SRES scenarios), and the results were input to the TAR.
- Climate models developed so far have been unable to deal with tropical cyclones, regional extreme events, heavy rains, etc. due to insufficient resolution. Impacts of global warming on these phenomena have been left in the TAR to be challenged in the next assessment report. These phenomena are the main targets in operational forecasts issued by the JMA for short (daily), medium (weekly) and seasonal ranges, all of which are based on the results of the numerical weather prediction (NWP). Therefore, the verification of predicted results by the NWP models is always made comparing them with observations. In order to facilitate the verification, JMA and MRI are now cooperating to develop a unified, basic and common model applicable to short range, weekly and seasonal forecasts and climate change projection.

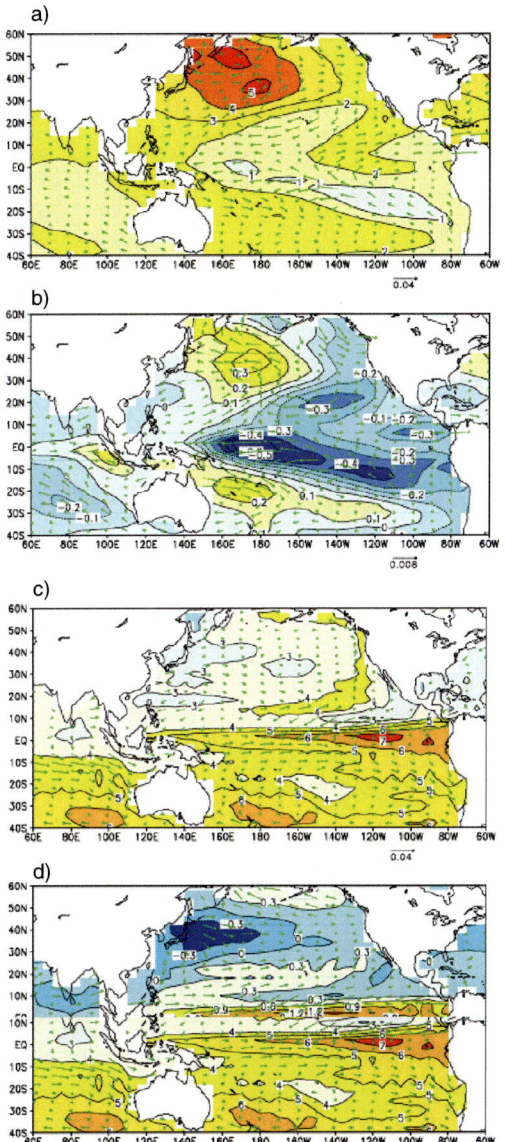

Figure 5. Comparison of spatial patterns between global warming and natural variability. a) Annual mean SST (contours) and wind stress (vectors) change for 1%/yr CO_2 increase around the time of CO_2-tripling (20-year mean over years 91 to 110) and b) annual mean SST and wind stress regressed onto the time series of EOF1 of the Pacific SST anomalies for the control run where the variance is normalized to be unity, simulated with the MRI coupled atmosphere-ocean GCM. c) The same as in a) but for CO_2 doubling, and d) the same as in b) but simulated with the MRI coupled atmosphere-slab ocean GCM. See Noda (2000) for a similar comparison using other AOGCMs compiled in the IPCC DDC.

Center for Climate System Research (CCSR) / National Institute for Environmental Studies (NIES)

The Center for Climate System Research (CCSR), the University of Tokyo has developed the Atmospheric General Circulation model (CCSR/NIES AGCM) and a climate model (AOGCM), and has conducted global warming prediction experiments jointly with NIES since its establishment in 1990.

In the first climate model, the AGCM was coupled to the OGCM developed at the Department of Geophysics in the University of Tokyo. Numerical experiments were conducted using this model for an instantaneous increase and a 1% increase per year of CO_2 (Abe, 1997). The results indicated that the globally averaged surface temperature increase is about 2°C. The spatial pattern of the surface temperature is similar to the results of other models. The change of the precipitation pattern resembles an ENSO-like pattern.

Next, a numerical experiment was conducted by introducing the aerosols in the climate model in order to investigate the direct effect of aerosols (Emori et al., 1999). In this experiment, the direct effect of aerosols was computed correctly based on direct computation instead of modifying a surface albedo, where an interaction between aerosols and a moisture field is considered. The experiment demonstrated that the direct effect is not so large and is overestimated in the method of modifying the surface albedo.

For IPCC TAR, a numerical experiment introducing the indirect effect was conducted (Nozawa et al., 2001). The temperature was expected to increase less because the indirect effect is included. However, the results were surprising -- the surface temperature increase was larger than expected.

This can be explained as follows. Several model improvements were introduced before this experiment. Subroutines for the planetary boundary layer and conversion rate from cloud droplet to rain droplet were changed to better represent an atmospheric moisture field. These modifications may have caused the increase of the surface temperature.

In order to confirm the effects of these modifications, several numerical experiments were conducted by using the AGCM with a mixed-layer ocean. The result is shown in Fig. 6. These modifications induced a 1°C difference in the doubling CO_2 experiment. The results of the other models are displayed in the same figure. Those values are distributed in the same range because each center uses different subroutines in the model. At present,

Climate Modeling and the Projection of Global Warming 73

uncertainty of this order in the temperature increase due to CO_2 doubling is unavoidable as we don't know which subroutine is correct.

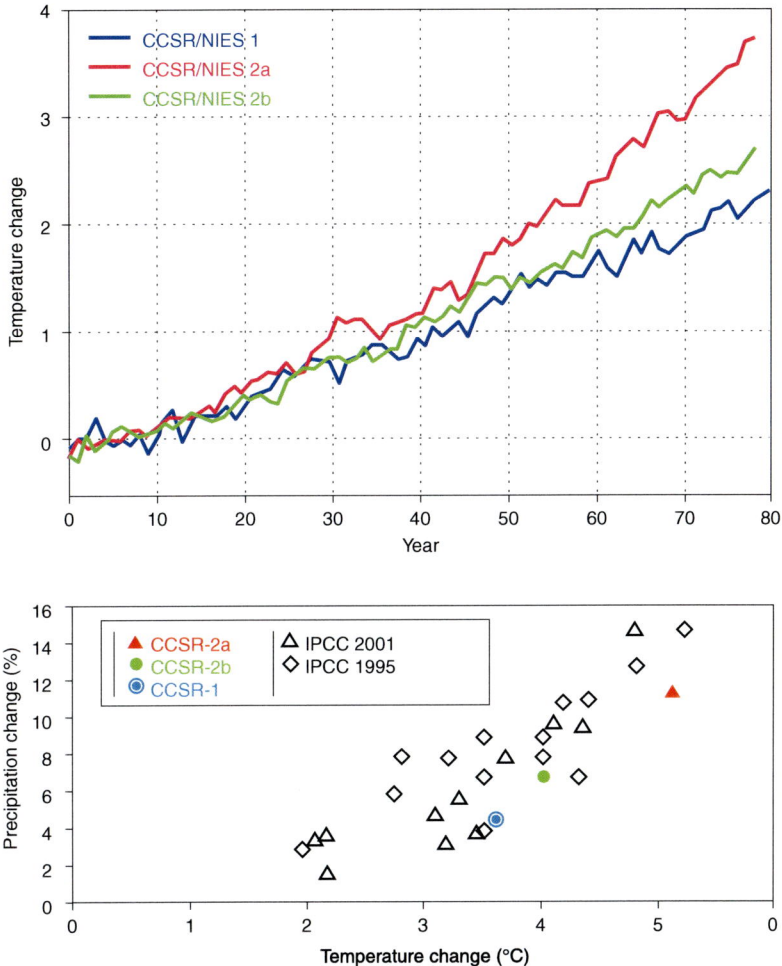

Figure 6. Top Change in surface air temperature for 1%/yr CO_2 increase simulated with the previous model (blue line), the model with the PBL routine replaced (green line), and the model with a new radiation table (red line). The same model shows this order of the difference in surface air temperature when subroutines are changed.
Bottom The model sensitivities of the above results are compared to those of the other models in IPCC (2001). The red, green and blue signs correspond to the same colors in the top panel (Sumi, 2001).

Frontier Research System for Global Change (FRSGC)

The Frontier Research System for Global Change (FRSGC) has studied various feedback effects of changes due to global warming in clouds and water vapor and has simulated effects of global warming on meteorological phenomena such as tropical cyclones, extra-tropical cyclones and so on.

The following presents an analysis of the feedback effect of clouds, the most difficult problem as pointed out by the IPCC since the First Assessment Report.

In 1990, the First Assessment Report of the IPCC (FAR) summarized, for the first time, results of simulations conducted by advanced national research institutes worldwide on the equilibrium climate under the doubled CO_2 concentration in the atmosphere. Surprisingly, the equilibrium changes of the global-averaged surface temperature due to CO_2 doubling are 1.9 to 5.2 °C, with a large scatter among models. This large scatter was not attributable to the greenhouse effect of CO_2 itself, but to the differences in feedback effects of changes in water vapor, clouds, snow cover, sea ice, etc. resulting from the greenhouse effect. Later studies have clarified that the feedback effect of clouds is the main reason for the large scatter.

The effects of clouds on radiation are complicated and difficult to analyze or predict because they differ between lower clouds and upper clouds, between solar radiation and infrared radiation, and so on. One of the most difficult problems has been how the "whiteness" (reflectivity of solar radiation) of clouds changes as global warming proceeds. The difference in approaches to this problem has caused the maximum and the minimum of the projected surface temperature change among models, as described in "The Skeptical Environmentalists" (Lomborg, 2001), which has provoked controversy on environment issues.

Resolving the problem is not an easy task, but we have a hint in observation: the average surface temperature in the summer (June, July and August) in the northern hemisphere is higher than that in the southern hemisphere, with a difference of up to 3°C. This is the same value as expected for CO_2 doubling. The above annual cycle is caused by the asymmetry in land distribution between the hemispheres. Assuming that this phenomenon, though different in nature from global warming, could be a proxy for global warming, Tsushima and Manabe (2001) analyzed cloud data from satellite observations and model results, analyzing the annual cycle of the cloud reflectivity of solar radiation for data from both the Earth Radiation Budget Experiment (ERBE) satellite and from simulations by three climate models.

Figure 7 depicts the result. In each of the cases (three simulated and one observed), the abscissa represents the monthly mean value of global-average surface temperature and the ordinate the cloud reflectivity of solar radiation per unit cloud amount, with the numbers near plotted points indicating months. Clearly, the observed reflectivity of solar radiation changes little during the annual cycle, although there is a very slight tendency of higher reflectivity for months with higher global-average surface temperatures and each of the three simulations describes the same tendency but with much more dependence.

The above results suggest that every model incorporates a similar series of physical cause-and-effect processes, that is, that a higher temperature would cause more water vapor concentration in the atmosphere, leading to the production of clouds with more cloud water amount (total amount of cloud droplets per unit volume), which means that clouds have higher reflectivities to solar radiation in warmer climates. However, the observed cloud reflectivity changes little during the annual cycle. These facts indicate that the models may overestimate the negative feedback effect (suppressing effect on warming) of clouds through their reflectivity. Nonetheless, we cannot exclude the possibility that the almost null tendency in observation might stem from a cancellation between cloud reflectivity and some other adverse but unknown effects.

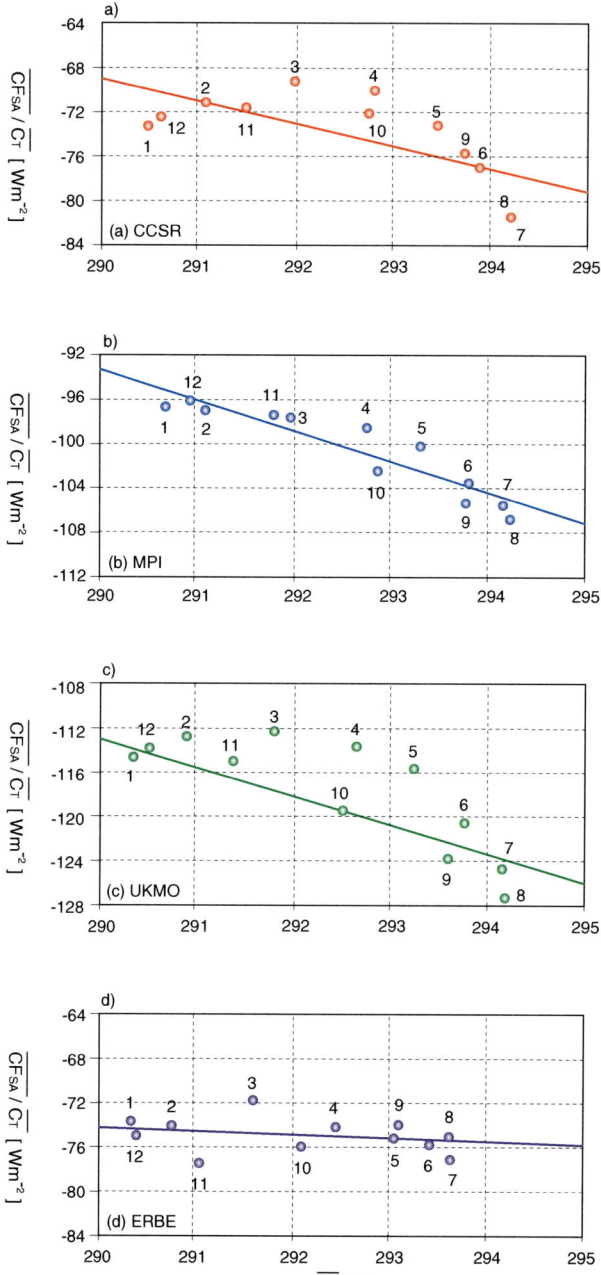

Figure 7. Globally averaged monthly mean values of annually normalized cloud radiative forcing per unit cloudy area (regarded as albedo) plotted against the global mean surface temperature in (a) CCSR/NIES model, (b) MPI model, (c) UKMO model, and (d) observation from ERBE (Tsushima and Manabe, 2001).

2.4.3 Research on Global Warming Projection by Regional Climate Models

Global climate models (AOGCMs) have rather coarse horizontal resolutions limited to about a 250km grid interval. Since such resolutions are inadequate for assessing regional climate change, the various methods listed below have been applied to extract regional-scale information (downscaling) from global model results.

1) Use of a high-resolution atmospheric general circulation model (AGCM) with the sea-surface temperature (SST) prescribed to the values predicted by an available AOGCM.
2) Use of a high-resolution regional model nested in the AOGCM, where the SST and the lateral boundary conditions are supplied by the AOGCM (see Fig. 8).
3) Use of long-term, empirical and statistical relations between meteorological quantities simulated, by an AOGCM and those observed in regional scales.

The resolutions of AGCMs used for Method 1) have so far been limited to a 100km horizontal grid size, while Method 3) involves an uncertainty as to how far we can apply the statistical relations found for natural variabilities to those for anthropogenic global warming. Therefore, Method 2) has been adopted to assess the regional impact of global warming in and around Japan as a more reliable downscaling method.

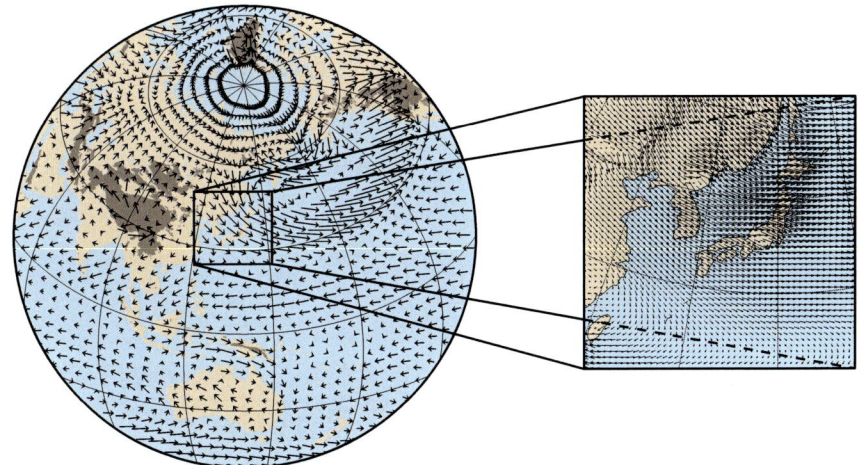

Figure 8. Regional climate model nested in a global climate model.

Downscaling Study with Regional Climate Models over and around Japan

Three Japanese research institutes (MRI, NIES, and Central Research Institute of Electric Power Industry (CRIEPI)) have been independently developing their own regional climate models to project the impact of global warming in Japan and its vicinity. Figure 9 depicts the results from global warming experiments by these three institutes under the same format. Experimental designs are different one another, as described below.

- MRI Model (see Sato, 2000): January climate is repeatedly calculated for 20 yeas by a Japan-area model that has a horizontal resolution of 40km and is nested in an Asia area model with a horizontal resolution of 120km, which is in turn nested in the MRI AOGCM with a horizontal resolution of 400km.
- NIES Model (see Emori *et al.*, 2000): Following the method described in 1) of the previous section, an AGCM with a horizontal resolution of 280km is first used for a 10-year run with SST set to the values obtained from the global-warming experiment with the CCSR/NIES AOGCM that has a horizontal resolution of about 560km. A Japan-area model is then nested in the above AGCM.
- CRIEPI Model (see Kato *et al.*, 2001): An East Asia area model with a horizontal resolution of 50km is nested in the NCAR climate model with a horizontal resolution of 280km and used for a 10-year run.

In Fig. 9, none of the AOGCMs used are able to sufficiently simulate characteristic climate features in winter over Japan, such as heavy precipitation (snow fall) along the side facing the Sea of Japan and dry weather along the side facing the Pacific. The pattern of precipitation is scattered among the models. However, these deficiencies found in the global models have been significantly improved and the scatter becomes much smaller for the simulations with the regional models (see the middle row in Fig. 9).

On large scales, most AOGCMs in general project the weakening of continental cold surges in winter from Siberia as a common feature of global warming. The resulting regional precipitation change is well captured along the Japan Sea side by all three regional models. However, a large scatter in precipitation over the Pacific side simulated by the AOGCMs is directly reflected in the projections with the regional models (see the bottom row in Fig. 9).

Thus the results of a regional climate model can be more effectively utilized if we know the predictability of the global model employed.

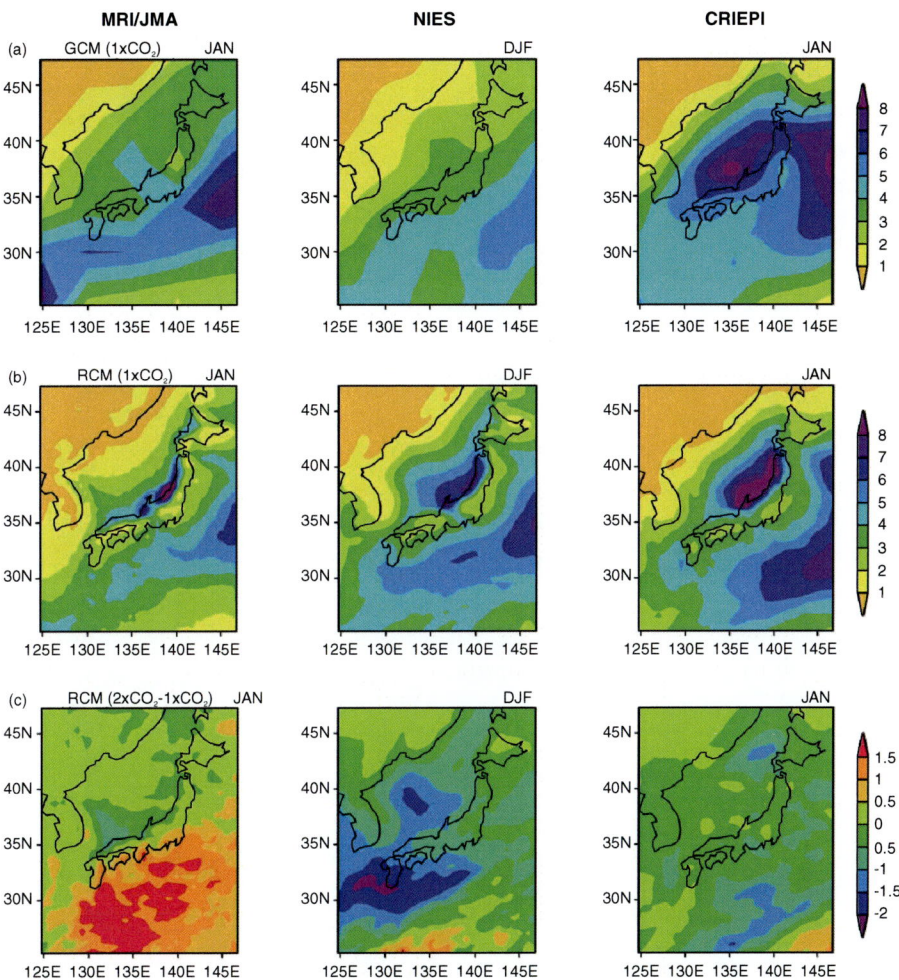

Figure 9. Daily mean precipitation (mm/day) simulated with AOGCMs (top), regional climate models (middle), and precipitation change due to CO_2 doubling with regional models (bottom). The results are from the models of the Meteorological Research Institute (MRI/JMA) for January mean (left), National Institute for Environmental Studies (NIES) for December-January-February mean (middle), and Central Research Institute of Electric Power Industry (CRIEPI) for January mean (right). The original data were provided by Y. Sato (MRI), S. Emori (NIES) and H. Kato (CRIEPI).

2.4.4 Climate Modeling in the New Era of the Earth Simulator

Future Directions of Modeling

Considering the progress in climate modeling, the present activities in climate change research, and strong social needs for increased reliability in global warming projection, the following are possible future directions of modeling under the new situation with rapid progress in advanced computational ability.

- Increase horizontal and vertical resolutions within tolerable limits in existing models. This would improve simulating relatively small-scale phenomena such as tropical cyclones that have been difficult to deal with.
- Develop new models with much higher horizontal and vertical resolutions beyond the limits of existing models. This would enable us to simulate severe storms or typhoons on much more certain physical bases and to address, in a more certain way, important issues including how the frequency of formation and the intensity of typhoons will change with global warming.
- Deal with model parameterization of physical processes in greater detail. For example, in calculating the reflection and absorption of solar radiation by clouds, we would now consider the size distribution of cloud droplets forming cloud layers, through tracing back to the formation of aerosols that are nuclei for condensation.
- Other challenges such as to develop a new integrated model for the Earth environment for simulating the carbon cycle and so on.

New Modeling Projects with the Use of the Earth Simulator

The number of modeling research groups in Japan capable of projecting global warming has been so limited that the groups have good communications with each other. They have discussed how to develop climate modeling for projecting global warming and extreme weather events. They have eventually come to share views on research objectives. In particular, they have come to almost common grounds towards the completion of the Earth Simulator with regard to how to develop models, what kind of experiments on global warming to conduct, and what kind of relevant roles could be shared among them.

In late fiscal year 2001, a new project called "Sustainable Co-existence Project on Human, Nature and the Earth" was designed by the Ministry of Education, Culture, Sports, Science and Technology (MEXT), to be launched in fiscal year 2002 as a big five-year research project to utilize the Earth Simulator. Most of the modeling research plans made at that time by researchers on a voluntary basis were materialized into this project. The following is a list of the main on-going modeling research subjects (including those under the above MEXT project) employing the Earth Simulator.

1) Development of a high-resolution climate model for future climate-change projection on the Earth Simulator.
   - In the first stage, simulation experiments of the 20th century climate and scenario-based projection experiments of the 21st century climate will be conducted, both by a coupled model consisting of an AGCM with a spectral horizontal resolution of T106 (roughly equivalent to 120km in grid size) and an OGCM with a horizontal resolution of $1/4°$ (in latitude) $\times$ $1/6°$ (in longitude).
   - In the second stage, model the coupling of an AGCM with a horizontal resolution of T213 (roughly equivalent to 60km in grid size) and an OGCM with a resolution of $0.1° \times 0.1°$, in coordination with the CCSR, the NIES and the FRSGC.
   - Cooperate with the Hadley Centre of the UK Meteorological Office to compare models for the same scenario-based experiments in order to further improve the models.
2) Global-warming experiment with a medium-resolution AOGCM in cooperation with the CRIEPI and the NCAR.
3) Development of super-high-resolution global and regional climate models.
   - Develop an AGCM with super-high resolution and a cloud resolvable regional atmospheric model. Both will utilize projected results from global-warming experiments with medium-resolution AOGCM. Using the future concentration of greenhouse gases and future sea surface temperature thus provided by the AOGCM (i.e., time-slice experiment), the AGCM will project changes in atmospheric phenomena such as tropical cyclones and Baiu fronts while the regional model will project changes in more localized extreme events such as severe rain storms.

- Presently, there is a plan to develop an AGCM with a resolution down to TL959 (roughly equivalent to 20km in grid size) and a cloud-resolvable, non-hydrostatic regional model with a resolution of several kilometers in grid size.
- This subject is being pursued by the JMA and the MRI.
4) Development of an integrated Earth system model for projecting Earth environment change.
    - Existing ordinary climate models are extended to include a carbon-cycle model, an atmospheric constituents model, and a biospheric model.
    - In the early stage, efforts are focused on conducting global-warming experiments taking account of the feedback through the carbon cycle. For this purpose, an AOGCM is coupled with a carbon cycle model covering biospheric interaction between the atmosphere, ocean and land.
    - This subject is being pursued by the FRSGC with the cooperation of the NIES, the CCSR and universities.
5) Development of a non-hydrostatic, super-high-resolution global atmosphere model resolvable meso-scale convective systems (FRSGC).

    The proposed model will have a horizontal mesh size of 5 km or less to eliminate the need for parameterization of cumulus convection, the most difficult of all the parameterizations. In order to attain high resolution, a new model will be developed with a "quasi-uniform grid" based on the icosahedral geodesic grid, instead of the spectral expansion or the latitude-longitude grid now being used for GCMs.

All the above subjects are consistent with the Global Warming Research Initiative. All of them except for 5) are now being implemented as subprojects of the Sustainable Co-existence Project on Human, Nature and the Earth.

NOTES

[*1] Jean Baptiste Joseph Fourier (1768~1830)
French mathematician and physicist. In his "Théorie Analytique de la Chaleur (1822)," he formulated thermal conduction as a boundary value problem of a partial differential equation, introducing the "Fourier series" and "Fourier integral."

[*2] John Tyndall (1820~1893)
British physicist. He studied the Tyndall phenomenon. He also made observations of the glaciers of the Alps. He is known as an illuminator of science as well as an Alpinist. His main work is "Glaciers of the Alps (1860)."

[*3] Svante Augustus Arrhenius (1859~1927)
Swedish physical chemist. In his paper, "Investigations on the galvanic conductivity of electrolytes (1884)," he discussed the electrolytic dissociation of electrolyte solution and proposed a new theory on acid-base, where a chemical substance producing hydrogen ions is an acid and one producing hydroxyl ions is a base. He also studied chemical kinetics, atmospheric electricity and cosmography.

[*4] Aerosols
A collection of solid or liquid airborne particles. There are various kinds of aerosols. For example, photochemical smog includes particles of sulfuric acid and organic materials. Aerosols decrease the amount of solar radiation reaching the Earth's surface by scattering solar radiation (direct effect). Aerosols also serve as condensation nuclei in forming clouds or modifying the optical properties and lifetime of clouds (indirect effect).

[*5] Lapse rate
The rate of decrease of an atmospheric quantity (usually temperature) with altitude. In general, temperature decreases with altitude in the troposphere. Lapse rate varies according to the site and the season with an average value of 0.65°C per 100m. When an air parcel is lifted adiabatically, its lapse rate is called the dry adiabatic lapse rate (about 1°C per 100m), if it is dry (without water vapor) or no condensation occurs. If the air parcel is saturated with water vapor, the lapse rate is called the moist adiabatic lapse rate, whose value depends on temperature and pressure (about 0.397°C per 100m at 25°C and 0.365°C per 100m near the Earth's surface).

[*6] Convective precipitation
Precipitation from cumulus convection. Cumulus convection serves as a heat engine in transporting the heat energy of the Earth's surface to the upper part of the troposphere. It is most active from the western Pacific areas to Indonesia.

[*7] Large-scale precipitation
Precipitation related to large-scale motions resolvable by the grid interval of a climate model. It is mainly associated with extra-tropical cyclones.

CHAPTER 3

IMPACTS AND RISKS OF GLOBAL WARMING

3.1 Global Impacts

3.1.1 Summary of IPCC Third Assessment Report

Climate changes caused by human activities include long-term temperature increases, changes in precipitation amounts and rainfall patterns, and rising sea levels, as well as short-term changes in the frequency and intensity of extreme climate events. All have serious impacts on natural ecosystems and human societies (IPCC, 2001; Ministry of the Environment, 2001).

The impacts of these climate changes differ among ecosystems and human societies, and estimating them requires appraisal of the interrelations between external forces, such as atmospheric temperature, precipitation amounts, and climate weather events, and the ability of natural ecosystems and human societies to resist or adapt to these impacts (Fig. 1). The vulnerability of an ecosystem or society to these external forces is determined by the relationship between its ability to resist and adapt. Given a consistent external force, vulnerability to climate changes will be higher in systems with a lower capacity to resist or adapt.

To assess climate systems and climate change, the IPCC Working Group I conducted simulations using a climate model (GCM: General Circulation Model) based on a new emissions scenario (SRES scenario), which predicted that by 2100, atmospheric temperatures will have risen to 1.4 to 5.8°C and sea levels 9 to 88 cm relative to 1990 figures.

Working Group II dealt with impact, adaptation, and vulnerability to climate change. The group used the above-predicted future values (called the general climate scenario) to forecast and assess impacts, but did not conduct studies using the high range (5 to 6°C) of the predicted temperatures. Studies based on these high atmospheric temperatures and sea levels are urgently needed.

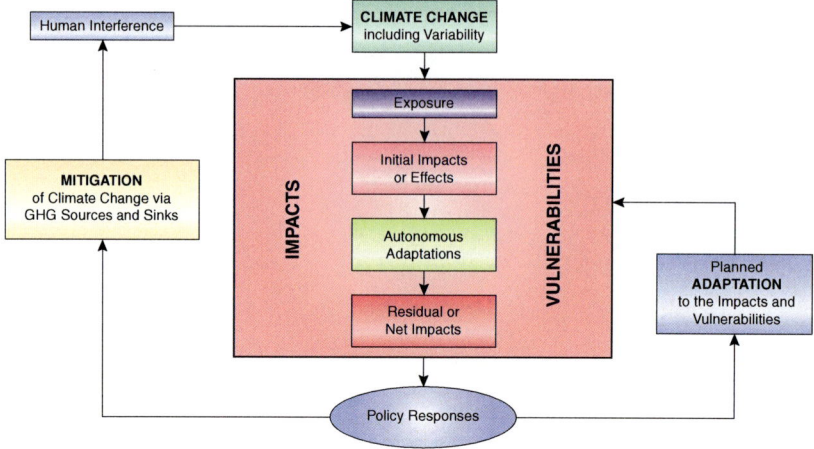

Figure 1. Scope of impact, adaptation, and vulnerability assessment (IPCC, 2001).

Glaciers, Sea Ice and Some Flora and Fauna are Already Affected

The IPCC Third Assessment Report concluded that there is new and stronger evidence that human activities caused nearly all the atmospheric warming seen over the past 50 years, and examined whether climate changes due to this atmospheric warming are already affecting vulnerable snow, ice and natural ecosystems.

More than 2500 research papers examining impacts on snow-ice and natural ecosystems were reviewed against the criteria of a focus on rising temperatures at the regional level and observational data recorded for more than 20 years. Of these papers, 44 regional and local studies were selected from among more than 400 dealing with plant and animal species, and 16 local studies from among about 100 dealing with snow and ice phenomena. The selected reports were then investigated in detail. Phenomena demonstrating a rise in temperatures were detected in the majority of these studies, from which it was concluded that the impacts of warming are already being felt.

Specific examples of confirmed changes are: retreating and shrinking glaciers, melting permafrost, late freezing and early thawing of rivers and lakes, longer growing periods in the mid and high latitudes, migration of plants and animals poleward and to higher altitudes, shrinking community and herd size of plants and animals, and earlier flowering of trees, emergence of insects, and hatching of bird eggs (Table 1).

Examples of the above that have been confirmed are mainly from North America, Europe, and the polar regions. There are almost no reports from Africa, Asia, or South America, as there have been virtually no studies in these regions.

Table 1. Examples of detected global warming impacts.

Detected impacts	Region	Phenomenon	Researcher
Retreat and shrink of glacier	European Alps	Decrease in volume of glaciers by 50% since 1850	IPCC, 1998
	Nepal	Rapid melting of Himalayan glacier. In 5 to 10 years, possibility of flooding and damage on local residents	UNEP, 2002
	Alaska	Melting of Alaska glacier	Arendt et al., 2002
Disintegration of ice shelf	Antarctic Peninsula	Disintegration of Larsen Ice Shelf (1995, 1300km^2; 1998, 300km^2; 2002, 3250km^2)	British Antarctic Survey US National Snow and Ice Data Center
Decrease in area and thickness of sea ice	Arctic Sea	Since 1950, thickness of sea ice in summer decreased by 40%, areas in spring and summer by 10~15%	IPCC, 2001
Delayed ice cover and earlier melting of river and lakes	Mid-high latitude in northern hemisphere	About 2 weeks shorter of ice cover period in rivers and lakes	IPCC, 2001
	Northern and central California	Since late 1940's, melting and runoff have shifted earlier	Dettinger and Cayan, 1995
Longer growing period of plants in mid and high latitude	Europe	Observed longer growing periods of 13 plants and bushes	Menzel and Fabian, 1999
Shift and movement Pole ward and higher place	Alps	In 10 years, alpine vegetation have shifted upward by 1~4m/year. Vegetation in mountain top have disappeared.	Grabherr et al., 1994
	North America	Shift northward of habitat of butterfly (Edith's Checkerspot butterfly)	Parmesan, 1996
	Australia	Increase in diameter of Norway spruce from 1961 to 1990	Hasenauer et al., 1999
Shrink and recovery of animal and vegetation communities	North Pacific Ocean, North Atlantic Ocean	Decrease in plankton concentration in North Pacific in summer by 30% since the 1980s, decrease in North Atlantic by 14%	NASA, 2002
	North-eastern US	Decline of red spruce	Hamburg and Cogbill, 1988
	Arizona	Reorganization of a semi-arid ecosystem, including increases in woody shrubs with increases in winter precipitation	Brown et al., 1997
	West African Sahel	Retraction of mesic species to areas of higher rainfall and lower temperature	Gonzalez, 2001
	Colorado	Reorganization of a shortgrass steppe ecosystem in a semi-arid site	Alward and Detling, 1999
Early flowering of plants, appearance of insects, and egg laying of birds	Southern Wisconsin	Phenological advances in flowering data in 10 herbaceous and tree species and no change in 26 over the periods 1936-1947 and 1976-1998	Bradley et al., 1999
	Europe and North America	During 1950-2000, 1 to 4 weeks earlier leaf unfolding, 1 to 2 weeks earlier leaf coloring, 1 week earlier flowering, 1 to 2 weeks earlier appearance of insects, flogs, and birds	Peñuelas and Fiella, 2001
	UK	4.5 days earlier flowering of plants in 1990s in average compared to 1954	Fitter and Fitter, 2002
	Japan	5 days earlier flowering of cherry trees for past 50 years	Masuda et al., 1999

Floods and Droughts are Increasing, and Impacts are Appearing in Some Countries and Regions

In recent years, extreme weather events such as floods and droughts have increased in some parts of Asia, so impacts on natural ecosystems and human societies are already beginning to appear.

The relation between climate changes and extreme climate events cannot be confirmed, however, because of limited observational data. Although they still have a high degree of uncertainty, simulations using climate models have suggested that climate change is related to typhoons and El Niño.

The effects of extreme climate events should be regarded seriously, since they have great impact and cause serious damage to natural ecosystems and human societies. Table 2 summarizes predicted extreme climate events and their impacts, listing the impacts of both simple extreme climate events, such as increases in maximum and minimum temperatures, and those of complex extreme climate events, including El Niño, Asian monsoon, and typhoons.

Moreover, climate change may cause large-scale, irreversible changes, which could produce impacts worldwide. Examples include slowing or stopping of major ocean currents (thermohaline circulation), large-scale melting of ice sheets in areas such as Greenland and western Antarctica, and release of carbon from massive withering and dying of land plants and thawing of permafrost. These changes are considered to have a very low likelihood of occurring in the 21st century, but accelerating climate changes may increase this.

Natural Systems and Humans are Sensitive and Vulnerable to Climate Changes

Ecosystems in nature are highly sensitive, with great susceptibility to external stresses and limited capacity to adapt. Such systems may therefore be unable to tolerate the effects of climate changes.

Impacts will be particularly severe on glaciers, coral reefs and atolls, mangrove forests, northern coniferous forests, tropical forests, the north and south poles, entire alpine ecosystems, and plains wetlands.

Areas of human society that will be most easily affected include all kinds of industries, the energy industry, human health and settlements.

It should be noted, however, that the effects of climate changes will not be entirely negative. Positive effects such as increased agricultural output and fewer people dying from exposure to cold in the early period of change will be experienced. Beyond this early period, however, the effects will be overwhelmingly negative.

Increasing Importance of Adaptation as well as Mitigation

Adaptation means making automatic or planned adjustments in one's behavior in response to a continuously changing climate. Humans act to mitigate the negative effects induced by climate change, and in some cases may be able to benefit from it. This requires expenditure, however, and it will not be possible to avoid all damage.

Natural ecosystems and human societies can, to a certain level, automatically adapt to climate changes. Human societies, moreover, can systematically devise and implement adaptive strategies. For example, in human activities such as crop cultivation that use climate as a resource, there is accumulated knowledge and experience for adapting to extreme climate events. This knowledge and experience may become the basis to apply adaptive measures to cope with climate changes forecast for the future.

While the importance of adaptive measures will continue to increase, there will be problems in implementing them in combination with measures to reduce greenhouse gases. For example, studies on methods to estimate the effects and costs of adaptive measures are lacking, and there are few case examples of their implementation. Studies that include economic assessment of the cost performance of adaptive measures, as well as comprehensive studies that combine this with measures to reduce underlying causes, will become increasingly necessary.

Table 2. Examples of impacts resulting from projected changes in extreme climate events (IPCC, 2001).

Projected Changes during the 21st Century in Extreme Climate Phenomena and their Likelihood[a]	Representative Examples of Projected Impacts[b] *(all high confidence of occurrence in some areas[c])*
Simple Extremes	
Higher maximum temperatures; more hot days and heat waves[d] over nearly all land areas (*very likely*[a])	• Increased incidence of death and serious illness in older age groups and urban poor • Increased heat stress in livestock and wildlife • Shift in tourist destinations • Increased risk of damage to a number of crops • Increased electric cooling demand and reduced energy supply reliability
Higher (increasing) minimum temperatures; fewer cold days, frost days, and cold waves[d] over nearly all land areas (*very likely*[a])	• Decreased cold-related human morbidity and mortality • Decreased risk of damage to a number of crops, and increased risk to others • Extended range and activity of some pest and disease vectors • Reduced heating energy demand
More intense precipitation events (*very likely*[a] over many areas)	• Increased flood, landside, avalanche and mudslide damage • Increased soil erosion • Increased flood runoff could increase recharge of some floodplain aquifers • Increased pressure on government and private flood insurance system and disaster relief
Complex Extremes	
Increased summer dying over most mid-latitude continental interiors and associated risk of drought (*likely*[a])	• Decreased crop yields • Increased damage to building foundations caused by ground shrinkage • Decreased water resource quantity and quality • Increased risk of forest fire
Increased in tropical cyclone peak wind intensities, mean and peak precipitation intensities (*likely*[a] over some areas)[e]	• Increased risk to human life, risk of infectious disease epidemics, and many other risks • Increased coastal erosion and damage to coastal buildings and infrastructure • Increased damage to coastal ecosystems such as coral reefs and mangroves
Intensified droughts and floods associated with El Niño events in many different regions (*likely*[a])	• Decreased agricultural and rangeland productivity in drought-and-flood-prone regions • Decreased hydro-power potential in drought-prone regions
Increased Asian summer monsoon precipitation variability (*likely*[a])	• Increased flood and drought magnitude and damages in temperature and tropical Asia
Increased intensity of mid-latitude storms (little agreement between current models)[d]	• Increased risk of human life and health • Increased property and infrastructure losses • Increased damage to coastal ecosystems

a Likelihood refers to judgmental estimates of confidence used by TAR WGI: very likely (90-99% chance); likely (66-90% chance). Unless otherwise stated, information on climate phenomena is taken from the Summary for Policymakers, TAR WGI.
b These impacts can be lessened by appropriate response measures.
c High confidence refers to probabilities between 67 and 95%.
d Information from TAR WGI, Technical Summary, Section F.5.
e Changes in regional distribution on tropical cyclones are possible but have not been established.

A Rise of More than 2 to 3°C in Mean Temperatures will Place a Great Burden on the Earth

Research on global warming has progressed from studies examining the entire world to recent studies examining specific regions or countries. Meanwhile, the movement to monitor the fluctuations of the Earth overall from each of these individual studies has become increasingly important.

Figure 2 presents the results of risk assessments of the impacts of rising temperatures. Five areas of concern have been identified (IPCC, 2001).

I. Unique and threatened systems like fragile ecosystems, glaciers, and sea ice
II. Extreme climate events
III. Distribution of impacts
IV. Global aggregate impacts
V. Large-scale, high-impact events

The impacts shown in white are neither positive nor negative, or the risk from them is insignificant. Those shown in yellow have a negative impact in certain regions or fields of activity, and red indicates major negative impacts over wide regions.

This figure demonstrates that there is a difference in how the impact risk emerges according to the extent of the temperature rise. For example, with regard to I and II, an impact is felt even if the rise in temperatures is small. With III and IV, there is no impact, or in some cases, a positive impact in certain systems or regions with a rise in temperature of up to 2 to 3°C.

Viewed from any number of perspectives, it is the developing countries, rather than the developed ones, that will sustain the greater impact. Moreover, since these developing countries are weak economically, the level of economic loss is greater as the atmospheric temperature rises.

In contrast, developed countries experience instances where a rise of up to 2 to 3°C will bring economic benefit, as well as instances where a loss will be suffered. All temperature rises beyond that will result in economic loss.

Finally, an aggregate calculation on a global scale indicates that, if the Earth's mean temperature rises 2 to 3°C, the world GDP will change plus or minus several percent. If the temperature rises beyond that, the loss will be greater.

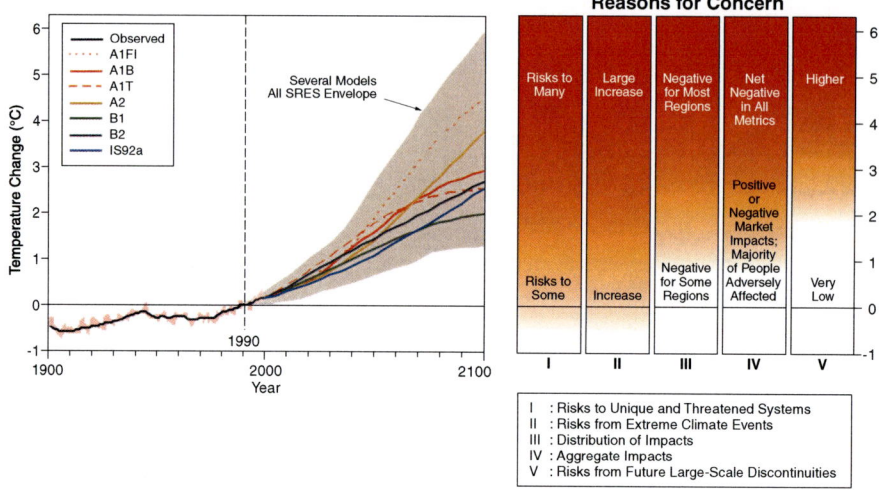

Figure 2. Reasons for concern about projected climate change impacts (IPCC, 2001).

3.1.2 Systems and Regions Vulnerable to Climate Changes

Impacts on Water Resources

Water is essential to the life of plants, animals, and humans. There are 1,386 billion km^3 of water on the surface of the Earth, of which 2.5% is fresh water. Of this fresh water, less than 1% can be used as water resources (Gleick, 2000). Climate changes are predicted to have a great impact on these meager, precious water resources.

An accurate determination of how much precipitation levels will change must be made to understand the extent of the impact climate changes will have on water resources. To make such estimates, climate models employing computers are used. Although there is some difference according to the model used, there is near consensus in predicting increased precipitation at high latitudes in Southeast Asia, and decreased precipitation in areas such as Central Asia, Mediterranean coastal regions, Southern Africa, and Australia. Figure 3 gives a sample forecast of river-flow volumes estimated using precipitation levels predicted with a climate model (CCSR/NIES) (Kainuma *et al.*, 2002). The features of the regions mentioned above can be identified from Fig. 3.

From this data, one can also understand the increasing risk of water

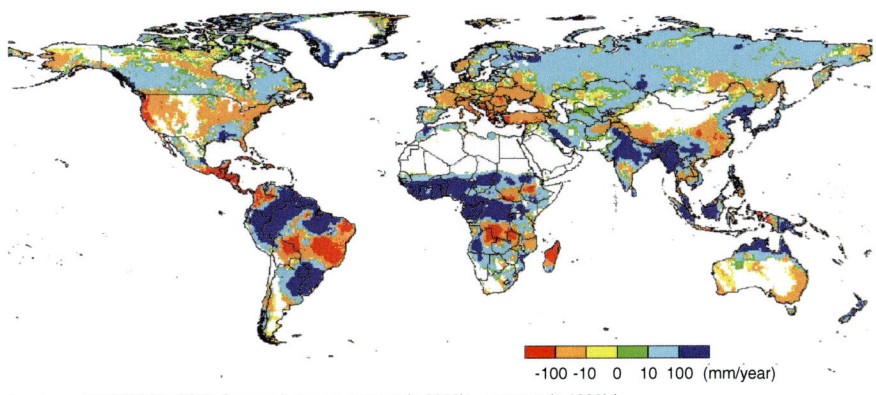

(mm/year, NIES/CCSR GCM, Change between average in 2050's ~ average in 1980's)

Figure 3. Predicted river flows after global warming using NIES/CCSR GCM (Change between average in 2050's – average in 1980's, Kainuma, *et al.*, 2002).

shortages. The use of more than 20% of the water resources that can normally be used by humans will result in water shortages. Judging from this, approximately 1.7 billion people in the world today are faced with the risk of water shortages.

Moreover, increasing populations or economic growth will result in increased water demand, so it is estimated that in 2025 the global population suffering water shortages will reach 5 billion. This will be especially acute in regions that currently suffer water shortages, such as Central Asia, southern Africa, and Mediterranean coastal countries. Conversely, in regions such as Southeast Asia, where precipitation levels are likely to increase, the magnitude and frequency of torrential rains and floods are expected to increase.

Furthermore, in regions dependent on snow for water resources, the winter snow will turn to rain, bringing the possibility that the peak flow volume of rivers will shift from spring to winter. This trend has already been confirmed in American rivers (NAST, 2001).

Impacts on Agriculture and Food Security

Securing food for a continually increasing world population is one of the major problems of modern society. The world's malnourished population, which stood at 840 million people in 1990, has shrunk to 777 million today (1997–1999), and by 2015 is predicted to decrease further by 200 million.

Meanwhile, in some developing countries, the dietary habits of people benefitting from economic growth are rapidly improving, thereby increasing demand for crop and feed to be used in meat production.

The IPCC Second Assessment Report released in 1995 (IPCC, 1996) stated that global warming would have some benefits for grain production, resulting in a sufficient food supply considering the increasing population. However, the Third Assessment Report reversed this optimistic prediction and indicated the possibility that the impact on agriculture would threaten food security.

We now consider what may happen in terms of the impacts on agricultural production and the food problem.

The agricultural industry uses climate as a resource, and has implemented various measures to adapt to changes. If the annual mean temperature were to rise 2 to 3°C, crop production would increase in the mid latitudes if systematic adaptive measures were implemented.

Conversely, damage from high temperature and water shortages will occur in the tropical regions, so that even slight increases in temperature will result in decreased production. In addition, food security will deteriorate in African and Asian regions due to extreme climate events such as droughts and floods, increased demand for food from growing populations, and rising grain and food prices on the international market.

Specific impacts on grain production will depend on such factors as species and cultivated variety, soil, disease and pests, fertilization effect of CO_2, atmospheric temperature, mineral nutrient content, interactions with air quality, and adaptive measures.

Global warming has advantageous aspects for grain production as well. One example is the fertilization effect of CO_2. However, while the fertilization effect due to increased carbon concentration was high in laboratory experiments, we understand that the effect is not as great in field experiments, and not so significant when placed among the negative impacts of extreme heat and drought. In the tropics in particular, damage from high temperature is likely to occur, disturbing grain production over wider areas.

Impacts on Terrestrial, Fresh Water, and Coastal Ecosystems

It has already been shown that the effects of global warming have appeared in natural ecosystems in every part of the world. Further impacts will likely emerge in the future in ecosystems with weak adaptive capacity, and these impacts will also affect plants and animals in other areas.

Terrestrial ecosystems have mechanisms for the absorption and release of carbon. Looking at total amounts for the past 20 years or so, we see that absorption has been greater than release. Because of this, it is thought that there is a relation between net carbon absorption and increased atmospheric concentrations of CO_2, rising atmospheric temperatures, increased plant productivity due to changes in soil moisture, and other changes.

However, there are predictions that, despite functioning as an absorption source until the mid 21st century, terrestrial ecosystems will become emission sources as a result of negative impacts from high temperature disturbances and other factors. The possibility of accelerated warming has been indicated if that turns out to be the case.

What about aquatic ecosystems? Cold-water fish such as salmon and trout will have decreased habitats, and in some cases may disappear entirely. However, fish that live in warm water will have expanded habitats, and it is predicted that the habitat borders of freshwater fish will shift poleward. Moreover, the deteriorated water quality in rivers and lakes due to eutrophication and other causes creates the possibility that the habitats of certain fish and shellfish will be threatened.

Coastal ecosystems such as coral reefs, saltwater marshes, and mangrove forests are diverse and productive. Impacts on these ecosystems include:
- Accelerated rise in sea level,
- The placement of obstacles such as dikes in open spaces to counter sea-level movements, and
- Rising sea-surface water temperatures and increased storms.

Further increases in pressures from both changes in climate and marine environments, as well as interactions with human activities in coastal environments, are therefore possible.

Bleaching of coral reefs has recently been confirmed in many parts of the world, and is thought to be due to the largest ever El Niño phenomenon, which occurred in 1997-98. The increase in sea-water temperatures as global warming progresses is certain to invite extinction due to more large-scale bleaching of coral reefs.

Human Health

The vector-borne infectious diseases of malaria and dengue are estimated to affect 40 to 50% of the world's population. These are very serious diseases in tropical and subtropical regions, and with climate changes the geographic range of potential infection will expand.

Whether infectious diseases occur and expand depends largely on pathogens and animal vectors such as mosquitoes; on having a sufficient human density within a single area; and on environmental conditions, socioeconomic factors, and the existence of public health facilities. Increased incidence of infectious diseases in developing countries in tropical and subtropical regions, and spread of these diseases to temperate regions, will be unavoidable.

Moreover, in cases when sanitary conditions deteriorate following rising temperatures or flooding, water-borne infectious diseases causing diarrhea and other symptoms may also spread. There is concern that the damage will be particularly severe in urban areas of developing countries with inadequate facilities for potable water supply or sewage treatment.

As global warming progresses, it may be presumed that the number of very hot days in summer will increase. As the increasing number of heat waves in which high temperatures continue for several days are compounded by factors such as rising humidity, the heat island phenomenon in cities, and air pollution, there will be an increase in the number of heat stroke victims and deaths. The impact of heat waves will be most keenly felt in cities, particularly among the elderly, sick, and those without air conditioning.

Although it has been said that in temperate countries the decrease in the number of deaths in winter will be greater than the increase in the number of deaths in summer, this would be limited to developed countries and cannot be generalized worldwide.

There will also be problems brought about by floods, with increased risk of drowning, diarrheal and respiratory illnesses, and hunger and malnutrition in developing countries. If tropical depressions, including typhoons, increase, destructive impacts are also predicted to occur in some low-lying coastal regions with high population densities.

Human Settlements, Energy, and Industry

Today 47% of the world's population lives in large cities located in coastal areas. The population shift from the countryside to urban areas is expected to continue in the future, so the impacts from rising sea levels and extreme weather events are expected to be greater in coastal regions in particular. Specifically, extreme increases in precipitation or rises in sea levels in coastal regions could cause flooding and landslides. Riverine and coastal settlements are therefore at particularly high risk.

Problems will occur more readily in huge cities of developing countries that do not have sufficient flood drainage, water supply, and waste management system capacity. Settlements where the population density is high and housing, safe water, and public health services are deficient, are extremely vulnerable to disaster. Regardless of whether a country is developed or developing, a rapid increase in population density due to rapid urbanization will also mean an increase in the value of assets exposed to extreme climate events. According to forecasts based on a mid-range scenario of a rise in sea level of about 40 cm by 2080, the average annual number of people suffering flood damage due to high tides in coastal areas will increase by 75 to 200 million people, and damage to facilities and assets is projected to be tens of billions of dollars in countries such as Egypt, Poland, and Vietnam.

The amount of damage suffered from disasters due to extreme climate events has increased rapidly over the past several decades. In the 1950s the global economic loss was $3.9 billion annually. This increased 10.3-fold to $40 billion annually in the 1990s (all in 1999 US dollars). The insurance covering such losses in the same period rose from near 0 to $9.2 billion. If the amount of damages grows beyond this level, it is possible that the private insurance system for such disasters will disappear, as the insurance industry will no longer be able to cope with the large amounts and will cut out the unprofitable departments or withdraw from the disaster insurance business altogether.

Regional Impacts

The IPCC Third Assessment Report divides the world into 10 regions, and comprehensively assesses the impact, adaptability, and vulnerability of each region based on the current state in the region, focused problems, and adaptive measures (IPCC, 2001).

For example, problems focused in the Asian region are ecosystems and biodiversity, agriculture and food security, hydrology and water resources, marine and coastal ecosystems, human health, and human society.

Negative impacts in the regions of Africa, Latin America, and Asia cover many fields, and include fluctuations in river flow volume, increased frequency of floods and droughts, decreased food security, smaller catches of fish, expanded health impacts, and loss of biodiversity. These will be a great impediment to future economic development.

North America, Australia, and New Zealand include groups such as indigenous peoples who have a low adaptive capacity. The adaptive capacity of ecosystems is also limited. In Europe, the southern and far northern (polar) regions are more vulnerable than elsewhere to impacts such as increased frequency of floods due to extreme climate events.

Climate change in polar regions will likely be large and rapid, and changes in ice sheets and glaciers, decreased area and thickness of sea ice, and deteriorated permafrost are likely. These changes will affect albedo, which may result in further acceleration of warming.

3.1.3 Front Line Impact Studies

Progress in International Impact Studies

Impact assessment of climate change is a relatively new field of study. The IPCC was established in 1988 and published a special report detailing technical guidelines in 1994 (IPCC, 1994). In the same period, the United States began a Country Study Project (USCSP) to investigate impacts and coping measures of individual countries, with consideration to developing the capacities of developing and economically transitional countries (USCSP, 1999).

The results of these efforts became the central findings for impact studies in developing and economically transitional countries of the IPCC Second Assessment Report. The Third Assessment Report focused on strengthening the adaptive measures of developing countries and promoting proper awareness of the strength of impacts. An international program to foster each region and country with the ability to examine these problems was initiated with Global Environment Facility (GEF) funding (Assessment of Impacts and Adaptations to Climate Change, AIACC).

In Asia, Japan took the lead in establishing survey projects to investigate impacts and countermeasures in developing countries.

Progress in Assessments of Impacts, Adaptation, and Vulnerability, and Future Issues

Progress in impact, adaptation, and vulnerability assessments over the past several years and future issues are presented below.

- Progress in studies on regional impacts of climate change. The collected data can now be used in individual countries or international projects.
- A methodology for investigating impacts has been established, and benchmark values determined. In addition, natural ecosystems and snow and ice that are vulnerable to climate change have been identified, and the need for monitoring them as indicators of change is increasing.
- There is heightened interest in extreme climate events because of the seriousness of the damage, but the number of studies remains insufficient. The amount of data on which to base future projections is small, and the views of researchers studying natural disasters should be sought in the future.
- The ultimate goal of the United Nations Framework Convention on Climate Change is to stabilize the concentration of greenhouse gases in the atmosphere, and progress is beginning to be made in some studies of emissions scenarios, climate models, and carbon circulation. One example shows estimates of impacts generated in the process of achieving stable concentrations (Fig. 4). This figure demonstrates that the production of wheat in India declines as stable concentrations are set higher.
- The fact that impacts are already being felt in various regions as warming progresses indicates the necessity of urgent investigations of adaptive measures. However, the stage has just been reached where adaptation has been defined, and frameworks and practical application of adaptation assessments are forthcoming issues.
- Research is progressing in the fields of natural ecosystems; agriculture, forestry, and fisheries; coastal regions; sea-level rise; human settlements; human health; and industry. There are relatively few studies on the impacts to human societies, and greater efforts are needed to collect additional data.

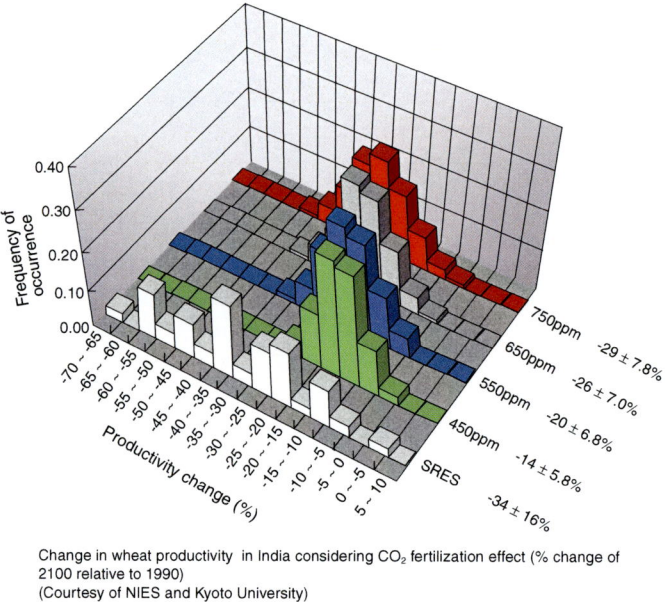

Change in wheat productivity in India considering CO_2 fertilization effect (% change of 2100 relative to 1990)
(Courtesy of NIES and Kyoto University)

Figure 4. An example of estimated crop impacts at several CO_2 stabilization levels (Courtesy of NIES and Kyoto University).

3.2 Impacts on Japan

In the early 1990s, Japan began actively conducting studies on the impacts and risks of global warming, the same when time international studies centered on the IPCC were proceeding. These studies covered a wide range of fields, including water resources and water environment; terrestrial ecosystems; agriculture, forestry and fisheries; marine environments; coastal zones; land preservation, disaster prevention, and people's lifestyles; industry and energy; and human health. Recently these studies have included estimates of impacts on developing countries in the Asia-Pacific region, and adaptive measures against these impacts.

The results of these studies and current knowledge regarding the impacts of warming on Japan (Harasawa and Nishioka (editors), 2001) are described below.

3.2.1 Emerging Impacts of Global Warming

Trend for Rising Mean Temperatures in Japan

There is a rising trend in the mean annual temperature in Japan, as evidenced by a rise of about 1.0°C over the past 100 years (see Fig. 2 in Chapter 1, Part 2). This rise in temperature began accelerating in the mid 1980s. Of the ten hottest years in the past century, eight were in the past decade, coinciding with the global trend. The rise in temperature in urban areas over the past 100 years has been more than 2°C, with that in Tokyo reaching 3°C. This large rise in the urban areas is partly due to the heat island phenomenon peculiar to cities. Even excluding this, however, Japan is clearly warming.

Changes in Organisms Sensitive to Warming

Living organisms and ecosystems detect warming and respond in various ways. In a worst-case scenario, a warming environment can lead a species to extinction.

Among the phenological observations conducted nationwide by the Meteorological Agency since 1953, the changes in the flowering date of the Japanese cherry (*Prunus yedoensis*) are particularly striking. These trees now flower 5 days earlier on average than they did 50 years ago (Fig. 5). In spring 2002, they blossomed at an unprecedented early date; holding cherry blossom festivals after the blossoms had already fallen to the ground was a first in people's memories.

There are a number of other examples of warming.
- Decreased alpine flora in Hokkaido, the north island in Japan.
- Expanded distribution of southern broad-leaved evergreen trees such as the Chinese Evergreen Oak.
- Nagasakiageha butterfly (*Papilio memnon thunbergii*), the northern border for which has been Kyushu and Shikoku Islands, appeared in Mie Prefecture in the 1990s.
- Appearance of the southern tent spider, seen only in western Japan in the 1970s, in the Kanto Region in the 1980s.
- Expansion of the wintering spot of the White-Fronted Goose to Hokkaido.
- Appearance of tropical fish in Osaka Bay.

- Shifting habitats of ermine and grouse on mountains such as Hakusan and Tateyama to higher elevations. There is some danger of complete disappearance.

These and other indications of diverse changes have been observed, and demonstrate that impacts of warming have begun to appear in organisms and ecosystems of Japan.

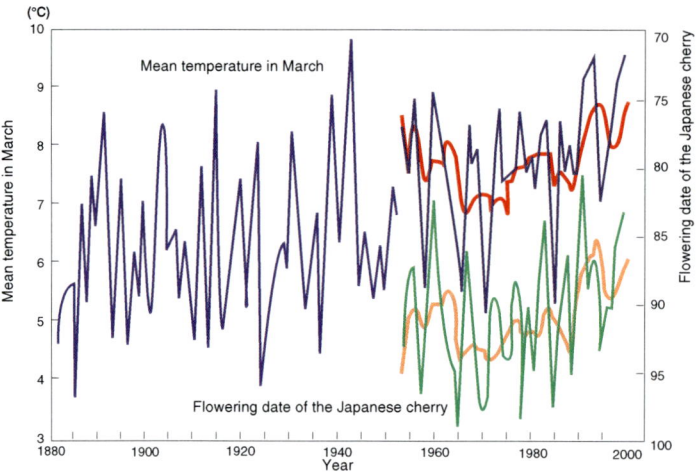

Figure 5. Annual changes in flowering date of the Japanese cherry (*Prunus yedoensis*) and mean temperature in March in Kyoto, Japan. Bold lines indicate five-year running means (after Dr. Keiko Masuda).

3.2.2 Impacts on Natural Environment

Impacts on Forests and Vegetation

Beech forests are typical of the cool temperate zone and are widely distributed in Japan. However, at the southern limit of their distribution, atmospheric warming will cause the transition of these forests into evergreen forests. Today, more than 40% of forests in Japan are artificial, but as the surface temperature increases, the environments where Japanese cedar and cypress have been planted are changing from beech zones to *shii* and *kashi* zones, two varieties of oak, and evergreen trees become competing species in these afforested lands.

The habitat range of large mammals such as deer, monkeys, and boar that live in forest zones is expanding, due to decreased snow accumulation volume and period due to climate change. If there is less snow, the survival rate of wild animals increases, as does the number of individuals. This results not only in damage to crops as these animals forage, but also more conflicts with humans.

Among the changes in potential vegetation in Japan, a great decline in the distribution of alpine vegetation and subalpine coniferous forests is predicted to occur by 2050. There is also risk that northern coniferous forests will change to broad-leaved deciduous forests, and that southern broad-leaved deciduous forests will change to broad-leaved evergreen forests, as shown in Fig. 6. This figure also shows a marked reduction in mountain ecosystems with a northward shift in the vegetation zone for the entire Japanese archipelago.

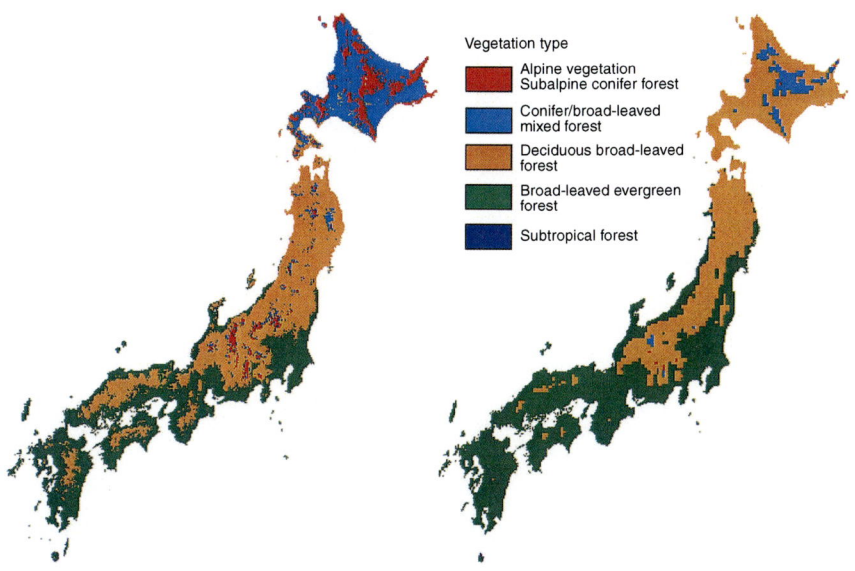

Left: Distribution of potential natural vegetation at present
Right: Distribution of potential natural vegetation for 2050 climate predicted by CCSR-98 scenario

Figure 6. Predicted changes in the distribution of potential natural vegetation between the present day and 2050 (Ishigami *et al.*, 2000).

Impacts on Biodiversity

Impacts from warming also include the risk of decreased biodiversity. Lands susceptible to impact include mountain and alpine regions, islands and isolated shorelines and beaches, and trees in urban areas. Since these ecosystems cover small areas, it may be difficult for them to maintain the numbers of individuals needed to prevent extinction.

Particularly at risk are organisms that live in a limited area. Examples include species with many varieties with mostly low survival capacities, such as *yakutanegoyo* in Japan, a variety of Armand's pine that lives only on Yakushima and Tanegashima Islands. Plant communities native to islands of the Southwest group or other small islands are highly susceptible to risks due to warming.

Deteriorating Water Environment and Fresh Water Ecosystems

In shallow lakes such as Kasumigaura, rising water temperatures and increased precipitation are thought to lead to higher chemical oxygen demand (COD) and concentration of nutrients such as nitrogen and phosphorous, which contribute to deteriorated water quality such as decreased transparency and eutrophication. In coastal waters as well, oxygen-deficient water bodies are more readily formed than in the past.

Moreover, as the sea-level rise causes further seawater invasion into estuaries, the salt concentration in many brackish lakes will increase, resulting in changes of ecosystems.

In river ecosystems, there is concern that fish habitats in cold water, such as Dolly Varden trout and white-spotted char, will greatly decrease. However, even if some species decrease, the diversity of the ecosystem would be maintained by the introduction of new species, although a decrease in the diversity of species will be unavoidable if this does not keep pace with the climate change.

Changing Ocean Environment

Recently, observations have demonstrated that the water temperature of the deep layer of the Japan Sea has been rising, and that a shrinking area of sea ice in the Sea of Okhotsk was caused by rising sea-surface temperatures. In addition, trends in sea-level changes over the past 100 years indicate elevated levels on the Sanriku coasts and the Pacific Ocean side, and lower levels along the Japan Sea coast. This is because the impacts of the global rise in sea level are superceded by rising and falling ground motion due to plate tectonics around Japan.

As sea-water temperature rises, plankton that had previously lived only in the tropical and semitropical regions have begun appearing in the seas around Japan. Southern planktons that were not previously seen have appeared around Japan, triggering harmful damage to cultivated oysters and other shellfish. In addition, the appearance of rivals may reduce sardine catches, diminishing the value of fishing grounds in coastal regions, as sardine is a major target for fishing.

Changes in ocean ecosystems also impact large mammals situated at the top of the food chain. Indeed, it is reported that the number of polar bears has been decreasing in the Arctic Ocean.

Coastal Land Features and Ecosystems

Coastal zones contain the habitats of organisms extremely vulnerable to climate change. One of these is coral reefs. Coral reefs grow upward at a rate of about 40 cm in 100 years in average. Therefore, if the future sea-level rise exceeds that rate, reefs will not be able to keep pace. Even more serious is the rising sea-water temperature. The optimum water temperature for coral reef growth is 18 to 28°C. If high water temperature of more than 30°C continues, the algae that coexist in the coral symbiotically will separate from the reef, and the coral becomes discolored and dies. This is called coral bleaching. Coral bleaching occurred on a large scale in all parts of the world after the El Niño/La Niña in 1997–98. If such phenomena occur more frequently in the future, it will likely cause serious damage to precious coral reef ecosystems (Fig. 7).

Another major problem is the erosion of sandy coastlines. While the main causes of erosion are a decreasing sediment supply and block of longshore sand transport, sea-level rise will accelerate beach erosion. If the sea level rises 30 cm, it is estimated that at least 56.6% of the sand beaches in all Japan will be eroded (Fig. 8). If the sea level rises 65 cm to 1 m, sand beach erosion will reach as much as 81.7% to 90.3%.

Tidelands, which support rich ecological communities, are no exception. Because tidelands are cut off from the hinterland by dikes or other structures, they cannot recede inland even if sea levels rise, and they are drowned. Therefore tidelands, which have an extremely gentle mean slope of 1/300, will lose an area 150 m wide with a rise in sea level of 50 cm. If such disappearance of tidelands continues, it is likely to have a huge impact on migratory birds such as snipes and plovers.

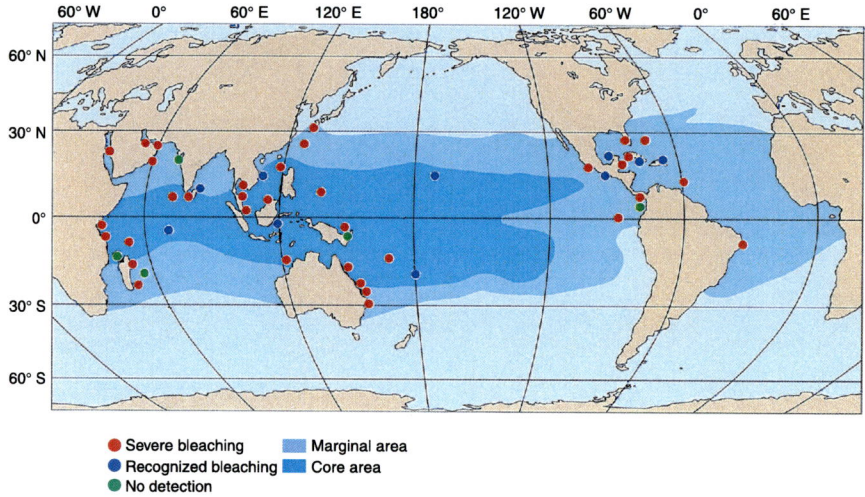

Figure 7. Coral bleaching during 1997 and 1998.

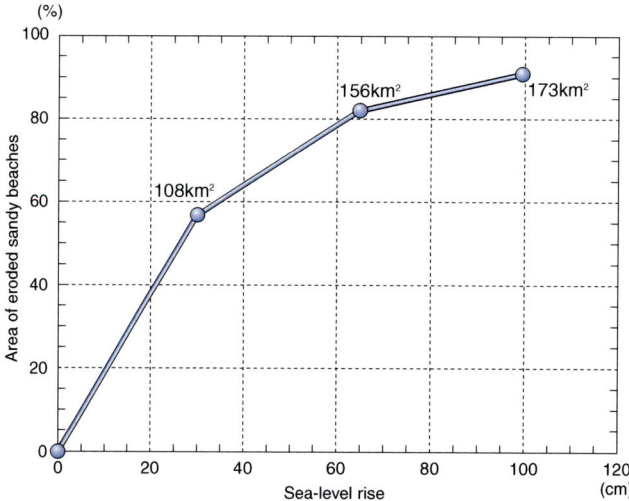

Figure 8. Predicted beach erosion due to sea-level rise (Mimura and Kawaguchi, 1996).

Considerations for Natural Adaptation

As described in the previous sections, the impacts of global warming are predicted to appear over a wide range of the natural environment. Whether natural environments can survive warmer conditions will depend on their adaptive capacity. Some environments will be very resilient against rising temperatures, changes in rain and snowfall, and rising sea levels, while others will be extremely vulnerable.

There will also be cases where the impact is further bolstered by human activities, as in the example of coastal tidelands or wetlands. These could survive by moving inland if sea level rises, but since most coastlines are covered by seawalls or dikes, they are blocked from receding. Ultimately, the areas of these landforms will shrink.

All adaptation of natural ecosystems to the warming will be reactive in nature. Therefore, to facilitate the natural adaptation we must provide natural systems with room to adapt. An example is creating spaces behind the present places in advance that can be used as evacuation corridors.

3.2.3 Impact on Human Society

Water Resources — Concerns of Flooding and Drought

The largest problems relating to water are flooding and drought. Recently, we hear news of intensified flooding in various regions of the world as well as drought reported in Africa and China.

In Japan, the frequency of extreme weather events of drought and flooding are increasing. There has been an upward trend in the last 100 years of incidences with very little annual rainfall (which generally occur once every ten years) across the country. The frequency of days with intensive rainfall is also increasing, which tends to cause severe urban flooding. These phenomena have a substantial impact on human society. For example, the 1994 drought in western Japan caused 11.76 million people to experience reduced pressure and restricted hours of water supply. Calculating the effect in simple economic terms, damage from the drought reached 140.9 billion yen or 1.2 billion US$.

The effect of climate changes will differ with areas depending on the extent of future rainfall change. One prediction for Japan indicates that flooding will be intensified, and that the river flow in snow regions will increase from January to March, and decrease from April to June.

Impact on Agriculture in Japan

In Japan, approximately 2 million ha of paddies provide about 10 million tons of rice each year, but a change in temperature would affect this productivity. Roughly, rice production will increase in high latitude and decrease in low latitude due to differences in growth and development efficiency. If the same cultivars are introduced in the future, it will be necessary to grow rice earlier in the Tohoku and Hokkaido regions, the northern parts of Japan, and later in the other regions to maintain current level of the yields.

Recent progress in research on the effects of carbon dioxide (CO_2) concentrations has revealed that a doubling the CO_2 condition would cause rice to ripe about 5% faster; and that over northern Japan, yields would increase by about 10 to 25% having no relation with variety of rice. While over middle and western Japan, the response would change depending on the variety of rice cultivated (Fig. 9). However, considering the general understanding, coupling with the problems caused by temperature rise, the impact would be predominantly negative. Further research in this area, therefore, is needed.

With regard to insect pests, a rise in winter temperature would push up potential wintering areas of pupa stage to the north, resulting in northward expansion of the habitats. It is predicted that the range and period of activity of insect pests would expand.

Impact on Food Security

As the Japanese diet has been westernized since the high economic growth period of the 1960s, the yield from domestic agricultural production continues to decrease with a rapid increase of food imports. As a result, the food self-sufficiency in calories has dropped to about 40%.

Japan depends in particular on imports from abroad for feed crops such as wheat and soybeans, which makes the country extremely vulnerable to impacts of climate change on the producing country. In 1999 the amount of imported wheat was 7 to 10 times the domestic production; those of soybeans and corn about 25 and 90 times their domestic production.

The major grain produced domestically is rice. Since irrigation facilities

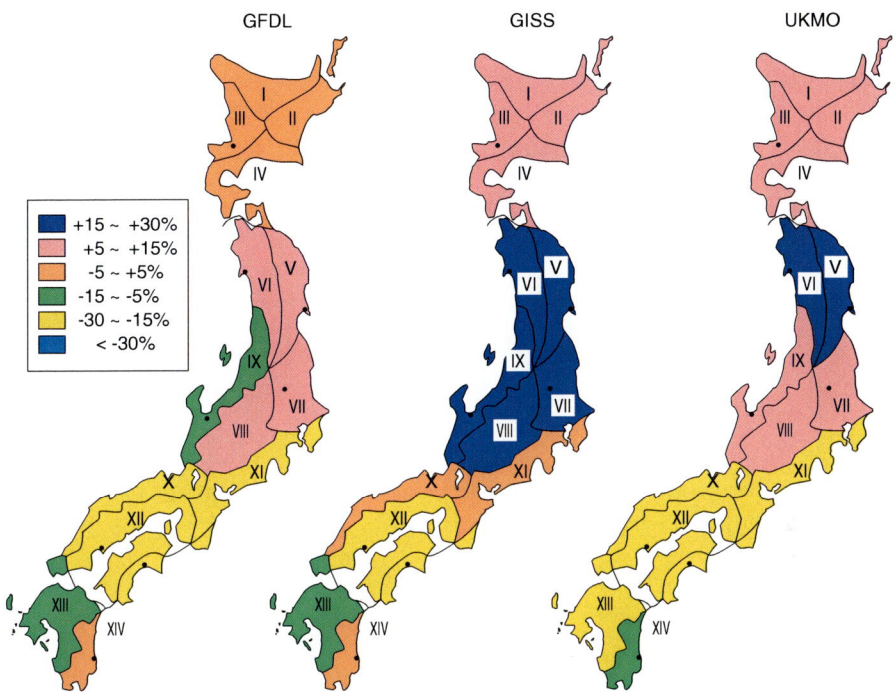

Figure 9. Effects of doubling CO_2 and increased temperature on rice yield (High temperature sensitive variety) (Horie *et al.*, 1995). These results were calculated based on the climate predictions given by Geophysical Fluid Dynamics Laboratory(GFDL), Goddard Institute for Space Studies(GISS), and United Kingdom Metrological Office(UKMO).

are fully developed, rice production is relatively robust to changes in climatic conditions. Regarding vegetables and fruits, as protected horticulture is widespread, negative impacts are not likely to threaten their stable production. Therefore, if food security in Japan is to be threatened in the future, it will probably be from plant disease or pests due to warming and frequent occurrence of unusual events including cold-weather damage.

It has been estimated that in the Asian region including Japan, the food supply will need to be increased up to twice present levels by 2050 to meet the increased demand by population growth and higher living standard. Climate change and sea-level rise may have serious effects on the food supply. We should recognize the possibility that the political and social problems would happen, if large-scale food instability occurs in Asian countries with enormous populations. To prepare for these problems, measures need to be implemented in both Japan and Asian countries to improve crop varieties for adaptation to warming, irrigation systems, and cultivation methods.

Coastal Disaster Prevention

Coastal regions are vital in terms of socioeconomic activity. Cities and towns facing the ocean account for 48% of the population, 48% of industrial shipment value, and 62% of commercial sales. Today two million people live in areas below the high-water level, with assets of 54 trillion yen. With a 1 m rise in sea level, these figures would more than double to 4.1 million people and 109 trillion yen.

In addition, sea-level rise would reduce the function and stability of disaster prevention facilities on the coasts. To maintain the function of seawalls and dikes at their current levels against a 1 m sea-level rise, seawalls should be raised 2.8 m on open sea coasts, and harbor quays raised 3.5 m in semi-enclosed bays.

Flooding and high waves are likely to affect the numerous other infrastructures on coasts as well, such as harbor and fishing port facilities, coastal roads, reclaimed land, pump stations, and drainage systems. Estimated costs required to prevent this are 7.9 trillion yen for harbor facilities and 3.6 trillion yen for neighboring coastal structures for a 1 m sea-level rise. Total expenditures would thus climb to 11.5 trillion yen.

Rising sea levels would also result in rising groundwater levels and increased salinity. This will weaken the supporting capacity of ground and make liquefaction more likely to occur during earthquakes. Because so much infrastructure and buildings are concentrated in the coastal areas with soft ground in Japan, the safety of cities against earthquakes would become a serious concern.

Investigations have begun to ensure the safety of coastal regions in the future (National Land Preservation Study Group for Sea-Levels Accompanying Global Warming, 2002). These studies include close monitoring of mean sea-level changes, and incorporation of the effects of future sea-level rises in the design of harbor and disaster prevention facilities. Internationally, adaptation strategies for coastal regions are classified into protection, accommodation, and planned retreat. The strategy thus does not rely solely on protection by disaster prevention facilities; investigations are also being conducted to change land use in regions at future risk, and to reduce vulnerability of coastal regions, including planned retreat.

Impacts on Industry and Energy

As global warming proceeds, human consumption patterns will also change, leading to changes in industrial structure. For example, if the mean temperature in June to August increases 1°C, consumption of summer products will increase about 5%, and if the period of high temperatures in summer lengthens, the consumption of air conditioning, beer, soft drinks, and frozen desserts will increase, so that electronics and food makers will likely need to reinforce their production systems for seasonal goods. However, it is still not possible to accurately forecast the extent of future effects.

Various impacts will also be felt in the supply and demand for electricity. Forty percent of the power demand in summer is for air conditioning, so a 1°C rise in temperature will cause an increase in power demand of approximately 5 million kW (amount for 1.6 million general households). In addition, the operating rate of factories producing summer products will increase, further increasing the demand for power (Fig. 10).

Changes in the amounts of rain and snowfall also have large impacts on hydroelectric power. The generation efficiency of thermal and nuclear power plants depends on the temperature of the cooling water, and a 1°C rise in coolant temperature will reduce the thermal power output 0.2 to 0.4%, and nuclear power output 1 to 2%.

Heightened Health Risks

Rising temperatures will have a direct impact on human health, with an increased overall death rate from heat stroke and other disorders. The elderly and people with underlying medical conditions will be at greatest risk.

Worsening atmospheric pollution and epidemics of vector-borne infectious diseases such as malaria and dengue are also possibilities. There have been recent reports of mosquitoes that transmit communicable diseases moving northward to the Tohoku region, and the risk of infectious disease may become a reality as the mosquito habitat expands.

However, social aspects play a greater role in the stress healthy people feel in daily life and work, as well as in the chronic diseases. Future research is needed with regard to the extent of the impact, the length of time, and how early it is likely to occur, and the regions where it is more likely to occur, given these social aspects.

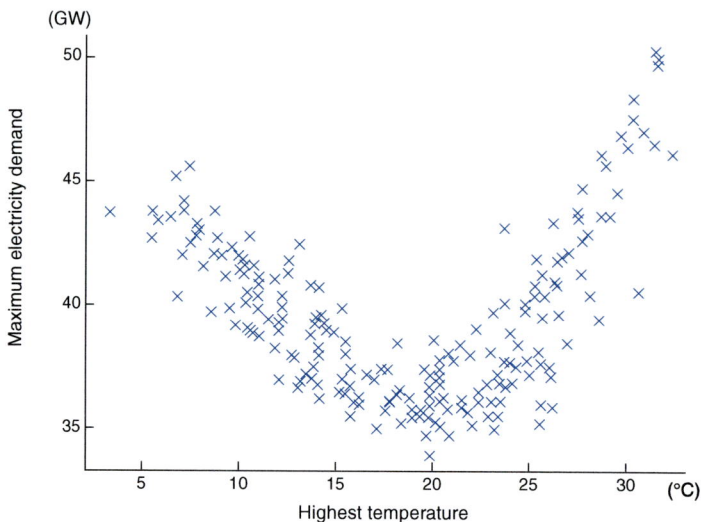

Figure 10. Relation between temperature and maximum electricity demand (Ohsaka *et al.*, 1996).

3.2.4 Status of Research and Adaptation
—What Kind of Information will be Needed?

As we have seen, research continues to clarify the impacts of global warming in an extremely broad range of areas. Table 3 lists the distribution of research activities to date. Numerous results have been obtained for terrestrial ecosystems; the agriculture, forestry, and fisheries industries; and coastal zones compared with other fields.

Table 3. Achievement map for impact and risk studies in Japan.

	Water resources Water environment	Terrestrial ecosystem	Agriculture, forestry and fishery	Ocean environment	Coastal zones	Land preservation, disaster prevention, and human settlement	Industry Energy	Human health
Impact detection		OOO		OO	O			O
Element studies on assessment methodology etc	OO	OOO	OOO	O	OOO	OO	O	OOO
National assessment Impact map	O	OOO	OOO		OOO	O		O
Threshold of impacts Vulnerable sectors and areas Economic assessment	O	OO	OO	OO	OO		O	OO
Adaptation	O		OO	O	O	O	O	O
Impacts on the Asia and Pacific region	O	OO	OO		OO			O

OOO : Studies with results in most areas   O : Studies in limited areas
OO : Studies with results in some areas   Blank : No studies or unapparent situation

In impact and risk studies, a wide range of research is needed, including detection of emerging impacts, impacts on individual sectors, nationwide assessments, identification of threshold of impacts and vulnerable areas, and adaptation strategy and measures. Many of the studies to date focused on fundamental aspects such as methods of predicting impact. However, to tie these with countermeasures against global warming, we need clear answers to the following questions.

- What extent (e.g., number of people at risk and monetary amount to be lost) will these impacts reach on a national scale?
- Which sectors in which regions will sustain the severest impacts?
- Threshold of impacts - How many degrees can the surface temperature rise and how many centimeters can sea levels rise before the world will have intolerable impacts?
- When will these occur?

Table 4 summarizes our current understanding of the critical values for impacts. Although we have obtained certain amounts of information, our knowledge remains insufficient to answer the fundamental questions above.

Measures against warming can be classified as either measures to mitigate global warming or those to adapt to a warmer world. Large efforts are clearly needed to prevent warming; however, we must also investigate adaptive measures to eliminate the deleterious effects of warming, as we cannot completely prevent warming by the current institutional and technical countermeasures. While improving the accuracy of impact forecasts, we must also investigate adaptive measures for severe impacts that will appear at an early stage.

Table 4. Threshold of impacts in vulnerable sectors.

Vulnerable Sector	Exposure System	Threshold
Ecosystem	Plants in high mountain Mangrove	Apparent effects for 2°C increase Cannot survive for 50cm SLR
Agriculture	Rice	Heat effect by over 35°C during flowering
Marine Ecosystem	Coral reef	Bleaching by 1-2°C increase in water temperature
Coastal Zone	Sandy beach Port and coastal structure	Erosion of 56.6% and 90.3% of sandy beaches by 30cm and 1.0m SLR 100 billion US$ for countermeasures against 1m sea-level rise
Human Health	Elder people	Increase of mortality rate for over 33-35°C of daily high temperature (regional dependence)
Economy	National economy Electricity	Negative effects for 2-3°C increase Demand increase of 5MkW for 1°C increase in summer

CHAPTER 4

ASSESSMENT OF GLOBAL WARMING RESPONSE POLICIES

4.1 Introduction

The area of research on "global warming response policy studies" should be promoted based entirely on actual policy needs and practical policy development. This should not be at the mercy of individual, short-term policy responses, but providing systematic learning and fundamental knowledge to form policies that must have a higher order of priority in the long-term perspective.

Full-fledged global warming response policy studies began in the United States in the mid-1980s when the Department of Energy initiated a carbon dioxide research project. In the latter half of that decade, the Environmental Protection Agency followed with further comprehensive studies on policy options for stabilizing the global climate. Against the background of an international heightening of political interest in this area and reflected by the establishment of the Intergovernmental Panel on Climate Change (IPCC), studies in the global warming response policy area began in European nations, Japan, and other countries.

Regarding global environmental studies in Japan, the Environment Agency (now the Ministry of the Environment) appropriated a budget for the Global Environment Research Fund in 1989. It initiated funding for study and prediction of climate change and its impact as well as for studies on assessing response policies and mitigation technologies in the global warming area. These study programs played a very important role in advancing Japan's subsequent global warming response policy studies to the international level.

The Ministry of Education, Culture, Sports, and Science and Technology (formerly the Ministry of Education) has promoted global warming response policy studies by allocations under the Grants-in-Aid for Scientific Research as well as the Research for the Future Program. Ministries and agencies closely related to the global warming mitigation area, such as the Ministry of Economy, Trade and Industry (formerly the Ministry of International Trade and Industry) have also begun to advance such studies, although only on a partial basis.

The Kyoto Protocol of December 1997 was a major step forward in international politics concerning global warming mitigation, further invigorating research activities in this area. The IPCC compiled its Third Assessment Report in 2001, and global warming response policy studies were systematically reviewed in the Working Group III Report. Proposals were included for developing new research activities in anticipation of the Fourth Assessment in 2007.

The main global worming response policy studies that have been implemented under these circumstances can be divided into the following six areas:
- studies on long-term emission scenarios [*1] and mitigation scenarios
- studies on mitigation costs
- studies on assessing technologies and ancillary benefits [*2]
- studies on policy design
- studies on policy decision processes and international negotiations
- studies on establishing mitigation targets.

This chapter provides an overview of research trends in the world and the present status of research in Japan in these six areas, and also describes new movements in the development of future research activities.

It should be noted that funding for global warming response policy studies in Japan is dispersed among various sources compared to other areas of global warming studies, and is closely related to the policy-making processes of individual ministries and agencies. The researchers involved in these studies come from various disciplines including economics, jurisprudence, policy science, and engineering. These factors make it difficult to comprehend the situation from an overall perspective. The present status of research compilation centers on literature by Japanese researchers cited in the latest Assessment Report (IPCC, 2001) concerning IPCC's global warming response policy studies. In the future, we must appeal to the relevant scientific societies for their cooperation in systematically investigating and understanding the actual conditions behind such studies.

4.2 Formulating Hundred-Year Scenarios

The extent of global warming will be greatly affected by the directions taken in the development of human society. The projected changes in energy use and land use, and the emission scenarios for greenhouse gases and sulfur oxides will greatly differ according to the direction of social development. Large margins of error also appear in global warming projections, and thus the scale of response policies will also vary.

Estimating emission scenarios based on all available scientific data is therefore a fundamental element of global warming response policies. Japan is a world leader in research on emission scenarios (Fujii, *et al.*, 1998; Jiang, *et al.*, 2000; Matsuoka, 2000; Matsuoka, *et al.*, 1995; Mori, 2000; Mori, *et al.*, 1998; Morita, *et al.*, 2000a; Morita, *et al.*, 2000b; Rana, *et al.*, 2000; Yamaji, *et al.*, 2000).

Most of the climate change projections made in the 1990s were premised on an emission scenario prepared by IPCC in 1992. However, this scenario, called IS92a (referring to case A in the 1992 reference scenarios prepared by IPCC), was only premised on one direction of social development. Moreover, various social changes since 1990 were not considered since the IS92a scenario was based on 1985 data.

The disintegration of the Soviet Union, rapid economic growth and introduction of free trade in the developing countries of Asia, and other factors led to major changes in emissions of greenhouse gases and sulfur oxides in the 1990s. There were strong criticisms, however, from developing countries that the scenarios themselves reflected a one-sided point of view held by the researchers in developed countries.

These issues were assessed in the IPCC Special Report of 1994 (Alcamo *et al.*, 1995) and formulation of new emission scenarios for greenhouse and other gases were recommended.

In response, IPCC organized a special project team and initiated the preparation of new emission scenarios (Nakicenovic *et al.*, 2000). IPCC's original role was to scientifically evaluate academic papers that had already been published, and organizing such an independent research project was an exceptional case. Emission scenarios are basic information required for scientific study of the global warming issue, so IPCC was called upon to provide such information and to implement this project.

Six research teams around the world prepared new emission scenarios over a period of three-and-a-half years, fully utilizing large-scale computer models. Two of these teams were Japanese. The resultant scenarios were called "SRES scenarios," an acronym of IPCC's "Special Report on Emission Scenarios."

The most conspicuous feature of the SRES scenarios is their wide range. More than 400 non-IPCC emission scenarios have been made into a database by a Japanese research team (Morita *et al*., 1998a; Morita *et al*., 1998b). Figure 1 shows the remarkably large range of these scenarios. The scenarios range from those showing a tenfold increase from the 1990 level by the end of the 21st century, to others showing a decrease to the minus level over the same period. Each research team prepared a total of 40 emission scenarios based on various development scenarios. The range of the SRES scenarios thus covers about 90% of the scenarios prepared in the past (see Fig. 1).

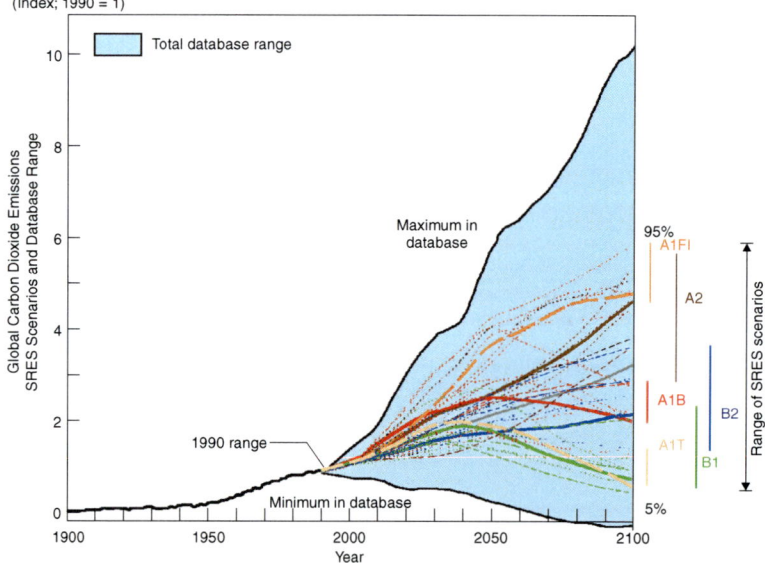

Figure 1. Global CO_2 emissions from energy and industry, historical development from 1900 to 1990 and in 40 SRES scenarios from 1990 to 2100, shown as an index (1990 = 1) (IPCC, 2001).

The SRES scenarios are based on six socio-economic development scenarios. Three scenarios, A1B (balanced rapid economic growth), A1FI (fossil fuel-intensive rapid economic growth world), and A1T (high-technology-oriented rapid economic growth world), are called the "rapid economic growth scenario group." In these scenarios, further economic growth is achieved worldwide, and there is great progress in education, technology, and social systems. The average economic growth rate over the past 100 years was about 3% annually, and these scenarios assume that this rate continues for the next 100 years, with the world average per capita income in 2050 exceeding 20,000 US dollars.

In the rapid economic growth scenarios, significant progress is made in technological innovation. Emissions of greenhouse and other gases greatly differ depending on the direction in which technological innovation progresses. It is for this reason that three scenarios were prepared, A1FI, A1T, and A1B. A1FI is notable for substantial innovations in technology, for the clean use of coal, and for innovations in oil and natural gas technologies. A1T projects substantial technological innovations in new energy including nuclear energy. In A1B these technological innovations are balanced.

There are three other scenarios in this group A2, B1, and B2. In A2 (heterogeneous world), each local region in the world values its own culture, and global economic and political blocs are formed by various social and political structures. B1 (recycling-oriented world) is based on a strong emphasis on the environment and society at large. Efforts are made for both environmental preservation as a global common asset and for economic growth on a global scale. All aim at balanced economic growth. B2 (local coexistence world) also places a strong emphasis on the environment and society. Rather than moving in the direction of focusing on global issues or solving problems internationally, however, too much importance is placed on local issues and equity, aiming at bottom-up growth (a management system in which decisions are made by proposals emanating from subordinate positions to higher ranks in the hierarchy).

Each scenario differs in terms of social-development direction and technological innovation; therefore, significant differences arise in economic growth, population growth, and energy use.

Climate change projections premised on the SRES scenarios have provided very important information from the scientific and policy standpoints. Various development directions are possible for humankind. The extent and significance of global warming and mitigation measures will differ greatly depending on those directions.

Figure 2 shows both the SRES scenarios and the mitigation scenarios to stabilize CO_2 concentrations that are based on them. In each case, the differences between the two scenarios represent the amount of emission reduction required (the length of the arrow). This shows that the greater the amount of reduction required, the greater the cost of mitigation measures and

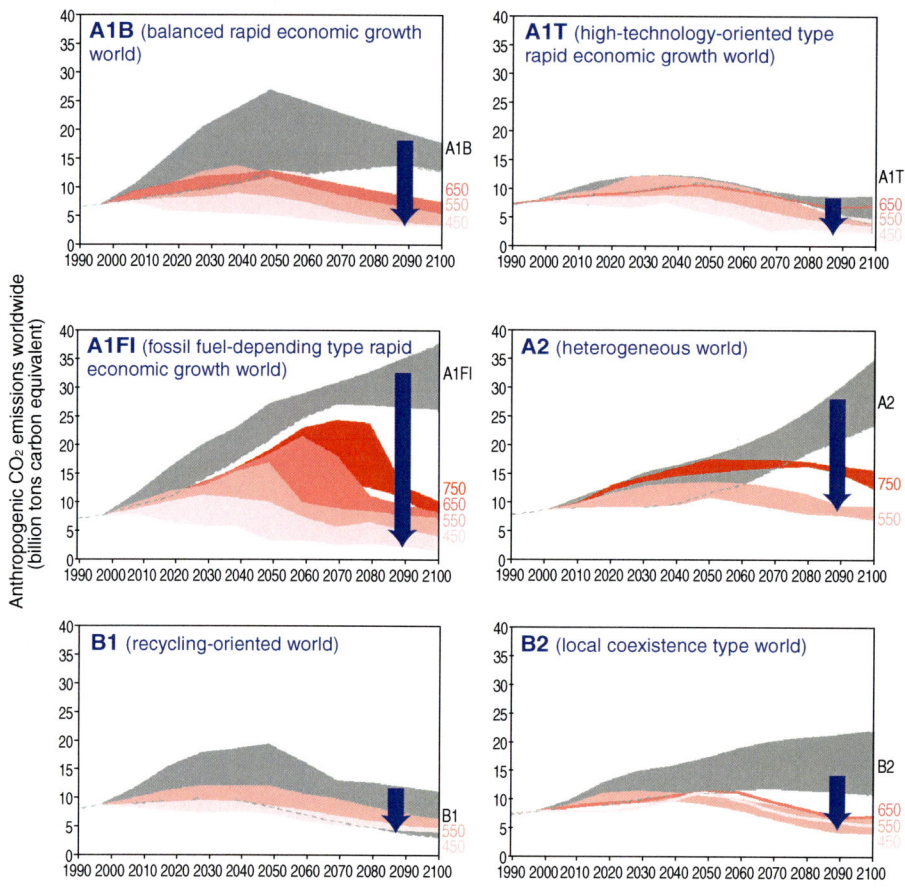

(The upper belt in each graph shows the emission scenario without mitigation measures, while the other belts show mitigation scenarios for stabilizing atmospheric CO_2 concentration at 450 to 750 ppm. The arrows show the reduction in emissions required for stabilization at 450 to 550 ppm. Each belt show the range of simulation results obtained by nine models.)

Figure 2. Differences in difficulty of implementing CO_2 reduction measures due to different paths for future development (based on IPCC 2001).

the greater the difficulty of implementing them.

Global warming mitigation measures become easy if development takes place in any of the three directions of the recycling-oriented world. In B1, recycling and efficient use of energy are pervasive. The local coexistence world (B2) is characterized by a reduction in waste and coexistence with nature. In the high-technology-oriented, rapid economic growth world (A1T), rapid growth continues through the introduction of nuclear energy, new energy, etc. It is only the society who can choose the direction in which it will advance.

Working Group III (IPCC, 2001) submitted the results of the above studies for inclusion in IPCC's Third Assessment Report. They drafted a chapter on mitigation scenarios based on the SRES scenarios. They also initiated studies on the relationships between the directions of socio-economic development and global warming mitigation measures.

These mitigation scenarios are called "post-SRES scenarios." Nine research teams throughout the world participated in their development, three of which were from Japan. The Japanese teams also coordinated the study program. Figure 2 shows their projected concentration-stabilization scenarios.

The current knowledge collated in the IPCC Report addresses two questions. Which variations in the development scenarios lead to which differences in the levels of mitigation measures taken and in the required technological innovations? Which measures are robust and which technological innovations are meaningful in whatever direction development advances?

These successive studies show that in order to cope with the enormous problem of global warming, studies on global warming response policies must not be narrow in focus. They must also integrate such policies with other socio-economic policies and development directions.

If the course of development creates affluence and conforms on the whole to the direction of global warming mitigation measures, such mitigation measures will also become a driving force for world development.

Scenario studies are currently shifting from the global scale to the regional and country levels. In the United Kingdom, each development scenario used as a basis for estimating global warming has been translated to the country level, and studies are progressing on how the effects of global warming will change according to the future situation of each region. Japan has also begun basic studies concerning the relationship between future development and global warming policy.

4.3   Estimated Costs of the Kyoto Protocol

IPCC's scenario studies demonstrate that the cost of long-term mitigation measures greatly depends on the direction of socioeconomic development. As far as cost studies are concerned, however, rather than dealing with the issue from such an extra-long-term viewpoint, attention is being focused on the issue over the next 10 years and toward achieving the targets of the Kyoto Protocol. As a result, we will need more accurate cost estimations and studies on cost reduction policies.

IPCC's Third Assessment Report published in 2001 mentions cost study assessments to achieve the numerical targets of the Kyoto Protocol, to include the Japanese studies (Kainuma *et al.*, 1999a; Kurosawa *et al.*, 1999).

Figure 3 depicts the range of marginal cost estimations by region, obtained by 11 studies that were assessed in this report. The median values shown are the middle values when all of the estimated values are put in descending order. In the case of an even number of data, the average value of the two numbers on either side of the middle is shown.

An examination of these costs reveals that when the Annex B (developed) countries [*3] implement the Kyoto Protocol without international emissions trading [*4] between Annex B countries, the marginal cost [*5] by 2010 is projected to be from 20 to 600 US dollars per ton of carbon equivalent ($/tC). The GDP loss is projected to be from 0.2% to 2%. Converted into the economic growth rate, the estimations give an annual rate of decrease of about 0.02 points to 0.2 points.

Japan's marginal cost is the highest, with the median at 300 US $/tC, and GDP loss is relatively small at 0.7%. The main reason for this is Japan's

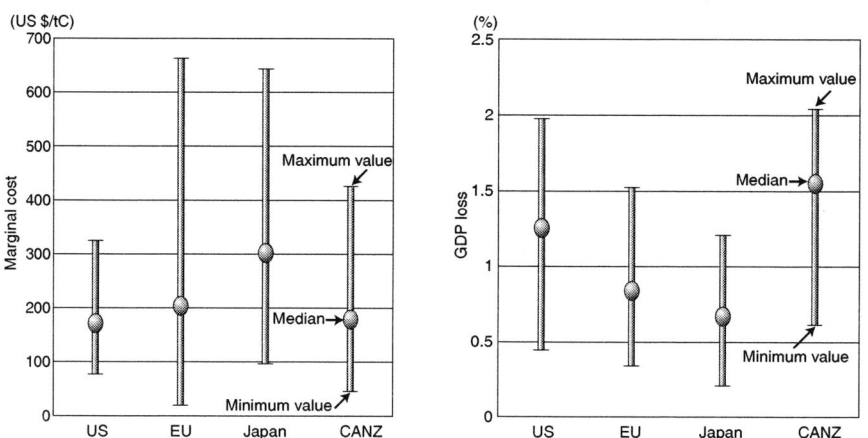

(In 2010, in the case of no emissions trading. Each set of values shows the range of simulation results obtained by 11 models. CANZ: Canada, Australia, and New Zealand.) (IPCC, 2001)

Figure 3. Estimation of marginal costs for achieving the Kyoto Protocol (left) and GDP losses (right) (IPCC, 2001).

lack of energy industries such as crude oil production. Conversely, the United States would experience a major effect on its energy industry, producing a relatively large GDP loss estimated at more than 1%.

Among Annex B countries, the effect on GDP in those undergoing economic transition, such as Russia, will range from a negligible increase to a rise of several percent. Non-Annex B developing countries, oil-producing countries may see a decrease in GDP due to the lowering of oil prices.

Additionally, the implementation of global warming mitigation measures by the Annex B countries will result in some carbon-intensive type industries relocating to developing countries with no obligation to reduce CO_2 emissions. Oil consumption will increase in the developing countries due to a decrease in the international trading prices of oil. This effect of a rise in CO_2 emissions in the developing countries (carbon leakage [*6]) has been estimated, with the increase projected to be about 5% to 20% of the CO_2 reduction achieved by the Annex B countries.

One promising means of reducing costs is utilization of the Kyoto Mechanisms [*7]. Costs may be greatly reduced by adopting a system to permit the transfer of emission rights between countries (international emissions trading). Further cost reduction will follow adoption of a system to enable developed countries to obtain emission credits for the reduction of greenhouse gases in return for investing in global warming mitigation programs in developing countries (Clean Development Mechanism [*8]).

When full-fledged emissions trading is carried out among Annex B countries, the report notes the projected marginal cost is 15 to 150 US $/tC and the GDP loss is 0.1% to 1%, half the level mentioned above. Since Japan has the highest marginal cost, it is expected to receive the largest cost-reduction benefit due to the Kyoto Mechanisms.

The costs of mitigation measures have conventionally been estimated by rough calculations using highly abstract economic models. By estimating GDP in the respective cases of implementing or not implementing a particular measure, this method identifies the macroeconomic loss resulting from its implementation. Such studies have been useful for understanding costs in general terms.

However, studies of practical policies for mitigation measures make it difficult to investigate detailed strategies for cost reduction using the conventional, highly abstract models. Consequently, there is a need for more accurate cost estimations that consider the reductions in costs resulting from the progress of mitigation technologies.

For this purpose, a "bottom-up" model that describes the processes and technologies of individual energy use in detail becomes necessary. The development of such detailed models is progressing, not only in developed countries, but in developing countries as well (Jiang et al., 1998; Kainuma et al., 1999b; Kainuma et al., 2000; Sato et al., 1999; Yamaji, 1998). Recent efforts have been made to develop models utilizing the characteristics of both by integrating the calculation results of these detailed models with those of highly abstract economic models.

Since the early 1990s, various research groups in Japan have estimated the costs of domestic measures for global warming mitigation and published various reports comparing these estimates. Initially, the range of the estimates was very large, with a disparity of more than fivefold in the projections for marginal cost.

In recent studies, however, this range has lessened. Without assuming dependence on the Kyoto Mechanisms, the estimate of marginal cost ranged from slightly more than 10,000 to about 35,000 yen/tC in order to achieve the targets of the Kyoto Protocol. GDP loss estimates ranged from 0.1% to 0.8% (Global Environment Subcommittee, Central Environment Council, 2001).

4.4 Technological Innovation Making Greater Progress Than Expected

There are several promising means of reducing costs other than utilization of the Kyoto Mechanisms. One of these is the development and diffusion of new technologies. IPCC's Third Assessment Report (IPCC, 2001) also recognizes that the success of global warming mitigation depends on future technological innovations and their dissemination.

The Second Assessment Report, released in 1995, systematically evaluated technology assessment studies. Technological improvements, however, have made more headway than expected. Wind power generation, technologies to suppress the generation of nitrous oxide and other greenhouse gases, hybrid cars [*9], fuel cell technology [*10], CO_2 underground storage technology [*11], and other recent technological advances have progressed at a speed that was unimaginable five years ago.

Table 1 provides an overview of the technological potential for reducing greenhouse gas emissions in 2010 and 2020. The categories with the highest potential by 2010 are energy-saving measures in people's daily lives

Table 1. Potential for emission reduction by 2020.

Category	Potential emission reduction in 2010 (million tons carbon, equiv./year)	Potential emission reduction in 2020 (million tons carbon, equiv./year)
Buildings	700 ~ 750	1000 ~ 1100
Transportation and Mobility	100 ~ 300	300 ~ 700
Industry - Energy efficiency improvement - Material efficiency improvement - Gases other than CO_2	300 ~ 500 ~ 200 ~ 100	700 ~ 900 ~ 600 ~ 100
Agriculture	150 ~ 300	350 ~ 750
Waste	~ 200	~ 200
Use of alternatives under the Montreal Protocol	~ 100	n.a.
Energy supply and source switchover	50 ~ 150	350 ~ 700
Total	1900 ~ 2600	3600 ~ 5050

(Note) Reduction potentials calculated based on technologies to be introduced in the market with a direct cost of 100 US dollars or less per ton carbon equivalent. The unit "tons carbon, equiv." means that emissions such as methane, nitrous oxide, etc. have been converted into CO_2 emissions based on their degree of contribution to global warming, and the total amount of greenhouse gas emissions is expressed by the weight of carbon. (based on IPCC 2001)

(households and offices) and in the industrial sector, with each achieving a reduction of as much as 700 million tC. In comparison, the effects of switching to new fuels and introducing new energy sources are relatively small, remaining at around the 100 million-ton level. This is because a 10-year period permits a limited chance for investment in switching plants and equipment to new fuels, and because of cost reductions through the introduction of new energy sources. By 2020, however, investment in plants and equipment and cost reductions will have progressed further, so the reduction potential will increase to a maximum of 700 million tons.

Greenhouse gas emissions by the transportation sector will rapidly increase by 2020 and their reduction will be a major task, but the reduction potential is estimated at a maximum of 700 million tons. In total, this will reach a maximum of 2.6 billion tons in 2010 and 5 billion tons in 2020.

In half of the reduction potential, the benefits produced by energy-saving exceed the cost, while the remaining half has the possibility of being achieved at a net cost (i.e., cost minus benefit) of 100 US \$/tC or less. Greenhouse gas emissions could thus be reduced to the 2000 level or below on a global scale during the period from 2010 to 2020.

There is a significant, long-term, technological potential after 2020 as well. Renewable energy resources such as biomass [*12], solar power, and wind power have very large technological potential, and in aggregate they could possibly satisfy all future energy demand. The enormous technological potential of carbon storage methods include CO_2 injection into disused gas fields and use of CO_2 injection for enhanced oil recovery. When the potentials of CO_2 injection into deep-sea areas and absorption by the terrestrial ecosystem are added, the figure far exceeds the total reduction of greenhouse gases necessary for the next 100 years.

The bottom-up model introduced above is exclusively used to estimate the effects of the development and diffusion of such technologies (Jiang et al., 1998; Kainuma et al., 1999b; Kainuma et al., 2000; Sato et al., 1999; Yamaji, 1998). The overall potential for emission reductions and the possibility of cost reductions have been estimated by describing in detail the processes and technologies of individual energy use and assuming the relationships between technological innovation and diffusion on the one hand and reductions in technology costs on the other. Various studies concerning individual technologies support these model studies. Japan is leading the world in research on energy-related technologies and carbon fixation technologies (Ishitani et al., 2000a; Ishitani et al., 2000b; Japan Resources Association, 1994; Kashiwagi et al., 1999; NEDO, 1999; Ogawa et al., 1998; Ozaki, 1997; Sato, 1999; Yagi et al., 1997). Studies on CO_2 absorption by forests have also recently begun in Japan.

Despite the presence of these technological potentials, the fact is that their diffusion has not been progressing as expected. Various obstacles hinder the spread of technology. For example, information on new technology is often not disseminated, and even when it is disseminated, management is often cautious about introducing such new technology. Additional barriers include a negative tendency toward investing in new technologies among banks and other financial institutions, and a lack of progress in new technology transfers to developing countries due to concerns over intellectual property rights. Many studies on the actual circumstances of such barriers and measures for overcoming them have been conducted around the world, although there have been very few studies of this type in Japan.

There is, however, another promising means of reducing CO_2 that should not be forgotten. This is the method of increasing ancillary benefits accompanying global warming response policies.

For example, enhancing energy efficiency in order to reduce CO_2 produces the ancillary benefit of reducing emissions of atmospheric pollutants. In developing countries, this benefit almost matches the cost of reducing CO_2 emissions. If so, the cost of CO_2 reduction in the developing countries could be regarded as virtually zero when including the benefit of avoiding the damage of atmospheric pollution.

Ancillary benefits can also be seen in the prevention of soil deterioration by afforestation carried out to increase CO_2 absorption, recovery of resources through recycling, and so on. UNEP [*13] and the US Environmental Protection Agency are implementing international projects concerning the ancillary benefits of global warming response policies, and Japan is gradually undertaking systematic studies as well.

4.5 Policy Design Growing in Sophistication

Global warming response policies must be designed in a way that skillfully combines and disseminates the individual technologies and mitigation measures described above, in order to reduce the processes involved and the costs borne by private enterprises as much as possible. Studies on global warming response policy design have made significant progress toward these goals over the past 10 years. The need has now arisen to formulate practical policies accompanying a succession of developments in this area. These policies include adoption of the Framework Convention on Climate Change at the Rio Summit in 1992, the First Conference of the Parties to the United Nations Framework Convention on Climate Change (COP1) [*14] in 1995, and agreement on the Kyoto Protocol at COP3 in 1997. General discussions on mitigation measures are almost complete, and major progress is being made in studies on specific measures as well as on combinations of measures.

In response to the need for practical policy design for global warming mitigation, studies and assessments are progressing on a menu of concrete measures on an individual basis. These include not only energy saving but also CO_2 reduction by recycling, absorption of CO_2 by afforestation and improvement of forestry management, reduction of methane and nitrous oxide emissions, and appropriate adaptation measures when global warming occurs.

Research is also being conducted on various cross-sector methods to implement these measures. For example, studies have begun on such policy measures as establishing performance standards to ensure the adoption of technologies with high cost-effectiveness, investing in research and development, voluntary efforts through agreements, consumer-side energy management, carbon taxes and emissions trading, and the Clean Development Mechanism.

In its report issued in 2001 (IPCC, 2001), IPCC emphasizes that for the global warming response policies of individual countries, we must determine the design of policy measures in line with the circumstances of each country. This must include regulatory measures such as energy-saving standards, and economic measures such as a carbon tax, domestic emissions trading, voluntary efforts through campaigns, and independent plans implemented by private enterprises. The report also points out that skillfully combining these measures (creating a policy mix) is the most efficient way of implementing them.

Japan is studying the design of various policy measures including:
- open forums based on the study of domestic systems,
- experimental studies and modeling of international emissions trading,
- studies on the design of systems for the Clean Development Mechanism, and
- various analyses concerning independent efforts by private enterprises (Aoyama, 1997; Fujime, 1998; Iwahashi, 1998; Japan Energy Conservation Center, 1997; Matsuo, 1998; Matsuo, 1999; Miyamoto, 1997; Yamasaki. et al., 1997; Institute for Global Environmental Strategies (IGES), 2000).

When it comes to research on policy mixes, Japan is lagging (IGES, 2002). For example, future studies are needed on how to link regional vitality and independent grassroots activities to global warming mitigation planning, and on how to integrate sustainable development of the developing countries with global warming response policies. Juristic studies are also urgently needed on consistency among systems when designing policy mixes.

Especially urgent policy tasks include the combination of environmental taxes, emissions trading, and independent efforts; the issue of taxation systems when taxes are exempted; the issue of consistency with existing taxation systems; and the issue of alignment between domestic and international trading systems.

In the area of policy design, there are very few studies in Japan that focus on how to deal with uncertainties in the global warming problem.

The most prevalent theory supporting the necessity for global warming response policies under conditions of uncertainty advocates the adoption of non-regret measures. In aiming at the implementation of global warming mitigation measures, even if global warming does not significantly progress, nothing will be lost if other problems are solved. This applies to cases in which ancillary benefits are produced, such as the promotion of energy saving in order to suppress CO_2 emissions, which will greatly contribute to the solution of resource problems even if global warming itself does not advance.

A slightly more positive argument for global warming mitigation measures under conditions of uncertainty is the need to cope with problems of uncertainty by flexibly combining policies. In this approach, various policies are first combined and then implemented with the aim of preventing global warming. Adaptive measures are then prepared in advance so that even if global warming does occur, the damage will be minimized.

Other suggestions are also being systematically examined, such as establishing some form of international insurance system as a preparatory measure against particular countries suffering major damage, acquiring the ability to adapt to crises in advance, and paying attention to ensure that dangerous situations are avoided. In the field of jurisprudence as well, theoretical studies on precautionary principle are being carried out to preclude dangerous policy choices.

4.6 New Insights for the Rules of Consensus

When assessing global warming response policies, it is important to assess both the value of the policy itself and the process of formulating the policy. Global warming response policies are formulated in a complicated milieu of conflicting international interests, and whether or not a policy can be skillfully implemented depends on the extent of cooperation from parties with conflicting interests. Consequently, if a wrong turn is made in the course of reaching consensus, the policy will not be realized or will not have much effect, even if its implementation is forced through.

Studies are therefore very important to analyze past political processes through which global warming response policies have been formulated, to determine what forms domestic policy decision processes should take (such as citizens' participation), and to clarify sound international negotiation processes. These studies have centered around economics. In such international studies, however, no such systematic investigations have been reported even by IPCC, and understanding the total picture is one of the tasks to be tackled in the future.

In the latest research trends in Japan, wide-ranging studies are being conducted on Japan's policy decision processes and there are high expectations for future developments in this area. These research trends can be summarized as follows.

Studies are progressing on the "Actor-based approach," in which analysis focuses on individuals and organizations that play important roles in policy decisions. Noteworthy studies in this area include those showing: that the role of nongovernmental organizations (NGOs) is expanding in global warming policy decisions; that the policy-decision coordinating power of individuals and organizations with medium power rather than those with great power is increasing; and that the formulation of policies through dialogue has become increasingly effective (Kanie, 2001).

Studies have been conducted to clarify the motives of individual countries in responding to the global warming issue, and to explore what forms effective international negotiation processes should take (Kawashima, 1997; Kawashima, 2000). According to these studies, the degree of importance each country places on the global warming issue in international negotiations depends on such factors as what degree of damage it will suffer as a result of global warming, what costs it will incur for global warming mitigation measures, what degree of leadership it can demonstrate in international politics concerning the global warming issue, and what domestic policy issues it has. The studies have also shown that the dynamics of negotiation can be explained by these factors.

Studies on the relationships between scientific uncertainty and policy decisions concerning global warming are also under way. In this research, confirming the importance of an integrated assessment process connecting science and policy decisions it is clear that the roles of new types of specialized work are becoming increasingly important for stimulating communication between science and policymaking.

The "environmental democracy" approach has recently attracted much attention (Amano, 2002). This is based on the recognition that for an issue like global warming that involves an enormous number of individuals and organizations, joint ownership of information on the environment by the people is essential, and people have the right to obtain information on the environment and to participate in policy decisions in order to secure an appropriate environment in which to live. From this position, studies are moving in the direction of fundamentally reforming the global warming response policy decision process and the international negotiation process.

4.7 The Difficulty of Judging the Balance of Losses and Gains from Mitigation Measures

What form should global warming response policies take after the Kyoto Protocol? What targets should be set and what should be the timing of response policy implementation? As a framework for advancing studies to answer these questions, IPCC's Third Assessment Report (IPCC, 2001) offers some highly suggestive analysis results. These are presented in Fig. 4.

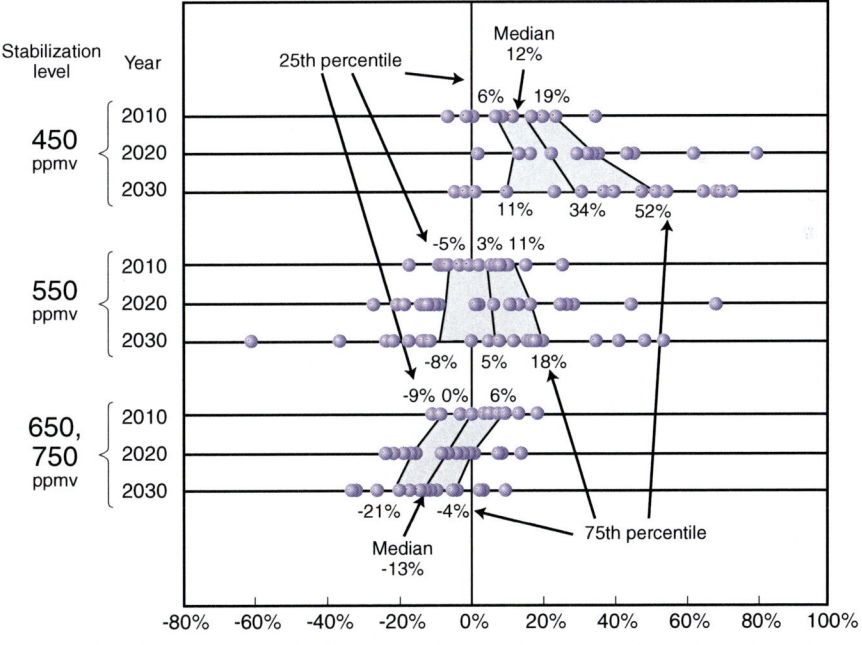

Figure 4. Required CO_2 reduction rates for Annex I countries shown by 76 stabilization scenarios (IPCC, 2001).

This figure shows the reductions in CO_2 emissions that need to be achieved by the developed countries from 2010 to 2030 in order to stabilize atmospheric CO_2 concentrations at 450 ppmv [*15], 550 ppmv, or 650 to 750 ppmv with respect to all 76 of the post-SRES scenarios mentioned earlier. Differences in the timing of emission reduction under each scenario and differences in the timing of participation by the developing countries cause a wide distribution in the amounts of reduction. However, if we focus on the middle half of the dispersed values, several important tendencies can be seen, including the following.

- The CO_2 reductions that must be achieved by the developed countries over the next 30 years greatly depend on the long-term stabilization target for atmospheric CO_2 concentration; that is, the targeted degree of global warming mitigation.
- If stabilization is targeted at 450 ppm, there is a high possibility that the reduction targets set for CO_2 emissions of developed countries after the term of application of the Kyoto Protocol will be stricter than those under the Protocol.
- If stabilization is targeted at 550 ppm, it is highly likely that there will be calls to set the reduction targets for 2010-2030 in the direction of maintaining the numerical targets of the Kyoto Protocol, on the premise that the United States also participates (approximately 5% reduction from the 1990 level by developed countries overall).
- If stabilization is targeted at 650 ppm or more, the numerical targets of the Kyoto Protocol are too strict. Therefore, after the term of application of the Protocol, there is a possibility that reduction targets could be eased.

These results may significantly change, depending on the future progress of technological innovations and new information concerning the carbon cycle. Of course, target setting is ultimately decided by political considerations. Consequently, they are meaningful only in providing a certain perspective.

It is clear that the setting of reduction targets after the term of the Kyoto Protocol will greatly depend on the target set for the mitigation of global warming over the long term. In order to determine greenhouse gas reduction targets over the next 30 years, the balance of losses and gains must consider both the benefits and costs of global warming mitigation measures over the period of the next century. Economic studies are therefore needed to demonstrate the extent of damage produced by global warming, or if it is reduced, the extent of benefits generated.

However, this balance of losses and gains with respect to response policies is very difficult to judge in the present circumstances, because studies to economically assess the impacts of global warming have not yet made substantial progress.

The accuracy of climate models has not reached the level necessary for economic assessments. Unless the accuracy of climate models is improved, they cannot be used to accurately estimate the extent of physical damage caused by global warming. There are some examples of studies by various countries, but they have not yet reached a level where they can be used for policy judgments. Some opinions say that global warming response policies should be postponed based on this situation, while many other opinions state the contrary.

One of these arguments arose from the assumption of technological innovations. This arises from the fact that most of the economic models assume that technological innovations will progress regardless of whether global warming response policies are introduced or not. If such models are used, technologies advance with the passage of time and technology costs decrease, so that postponing mitigation measures is less expensive.

In actuality, however, it is highly likely that by initiating global warming mitigation measures at an early stage, the timing of investments in technological development will be early, technological innovations will progress, and costs will decrease. Even so, these relationships have not been incorporated into the calculations because we cannot express in formulas the degree to which technological innovation will be advanced by investment nor to what extent a lowering of technology costs can be expected. Verification analysis of technological innovation is therefore an important future study theme.

There are also concerns that delay of mitigation measures may accelerate the pace of global warming. This point has been a major issue when deciding the timing of specific measures. The significant difference in the viewpoints of economics and ecology remain in the background of these concerns.

From an economic standpoint, the desirable form of development is maximizing affluence while maintaining assets and capital stock that produce income. In the ecological view, the desirable form is based on maintaining the recovery power and adaptability of natural systems that constitute a pivotal point, while enabling the stable workings of nature to continue. From this ecological standpoint, to maintain the recovery power of nature's pivotal systems, it is better to initiate mitigation measures as early as possible to lessen the speed of global warming. In order to maintain the workings of nature in a stable manner, studies to estimate the limits of emissions that nature can withstand have drawn a desirable picture of development from the economic standpoint within these limits.

In addition to the above aspects, there are opinions on the timing of mitigation measures from the standpoint of intergenerational equity. This issue arose from the question of how the costs of measures should be shared among generations. If we wish to lighten the burden on future generations, the timing of the measures should be set on the early side.

Presently, as we approach the conclusion of the Kyoto Protocol, these arguments and studies are about to be reactivated for the next step of the Protocol, discussions on setting targets for the second commitment period.

In the final section below, we will examine the future outlook for global warming response policy studies.

4.8 Response Policy Studies Entering a New Phase

The first point to be noted here is that the need for studies in the social sciences and humanities is increasing, encompassing not only economics but also political science, jurisprudence, sociology, and social anthropology.

Reviews of all the literature in the social sciences and humanities conducted by IPCC have inclined toward assessments from the standpoint of economics. However, in order to judge the soundness of global warming response policies and design more effective systems, contributions from the fields of social sciences and humanities have to be utilized as well. This was a major issue at the planning stage of the IPCC's Third Assessment Report. However, the stratum of global warming response policy studies in areas other than economics is very thin, revealing the fact that the reality does not fully correspond to the needs that exist. Further activation of studies in these areas is therefore necessary.

Interdisciplinary studies in the social sciences and humanities will become increasingly important in the future, with the progress of international negotiations such as the "law and economics approach" that analyzes the relationships between the mechanisms of obtaining international agreements and economic measures by integrating the fields of jurisprudence and economics.

Second, it is also necessary to bridge the gaps in studies between the developed countries on the one hand, and the developing countries and countries in economic transition on the other.

Since most of the existing policy studies have centered around the developed countries, large gaps arise when attempting to directly apply the previously developed models and policy designs to the developing countries and countries in economic transition. (In the IPCC Tokyo Workshop [*16] held in 1997, 12 serious gaps were found between integrated assessment models and the present situation in the developing countries.)

Measures taken by the developing countries do not hold the only key to solving the global warming problem over the long term. Policy agreements among the developed countries also depend on the creation of cooperative relationships with the developing countries and countries in economic transition. To achieve this, no time should be lost in eliminating these gaps. It is therefore necessary to advance regional studies of the developing countries and countries in economic transition.

In particular, mitigation studies targeting the developing countries in the future can be expected to place more emphasis on the characteristic systems of the region concerned and the values underlying them, coordination of interests between North and South, and equitable resource distribution among the developing countries. Consequently systematic efforts are required by researchers in the political science, cultural anthropology, and regional science fields.

Third, ratification of the Kyoto Protocol necessitates various urgent studies, such as the design and calculation of the effects of emissions trading (ET), joint implementation (JI), and the Clean Development Mechanism (CDM).

Studies on policy mixes, on policies for the expansion of carbon-absorbing sources, and on the possibility of reducing greenhouse gases other than CO_2 as well as related cost assessments are now urgent administrative tasks.

The need for these new studies must be treated not merely as a short-term administrative requirement. Rather, study plans should be formulated based on a firm, long-term research perspective.

Fourth, moves to explore global warming response policies after 2010 have become active. These efforts involve examining what forms global warming response policies should take in order to achieve the objective defined in Article 2 of the Framework Convention on Climate Change [*17], by studying global warming response policies from the perspective of other global policy frameworks such as international politics, trade, and science and technology.

The heated arguments seen in IPCC and elsewhere concerning equity between the developing and developed countries have been rekindled over the participation of the developing countries in the second commitment period and the setting of targets for the next period. Such problems as the friction between North and South, international industrial competition centering around the environment by the Japan-Europe environmental axis, and re-participation of the United States in the framework of the Kyoto Protocol also require analysis from the broader perspective of global politics. Discussions to integrate global warming response policies and other global environmental policies, and monitoring and assessment of the status of implementation of policies under the Kyoto Protocol, will be directly reflected in the actual policy processes of each country and linked to discussions on the second commitment period.

In Japan, examination of the policy factors for achieving the targets of the Kyoto Protocol has almost been completed. The following are future policy themes:
- compatible combination of policy factors taking international frameworks into consideration,
- quantification of policy effects toward the policy review at the end of fiscal year 2004,
- control of emissions in people's daily lives (household and office sectors) and in the transportation sector, and
- effects of cooperation with developing countries centering around a combination of CDM and official development assistance (ODA), as well as technical transfers and assistance for building up capabilities.

Equally important are verification studies concerning how to strengthen industrial competitiveness under the world order of new global warming response policies.

Finally, when conducting global warming response studies in Japan, we must consider the extent to which Japanese studies have been able to contribute to the world. Policy studies in which Japan has a high profile should therefore be especially encouraged, and the role of such studies in global study activities should always be clear. In particular, promoting joint studies with the developing countries of Asia and the Pacific, and contributing to international cooperation and understanding, should be considered.

NOTES

[*1] Long-term emission scenarios
Scenarios that quantitatively estimate future increases in aerosol and greenhouse gas emissions based on projected socio-economic developments such as population change, economic growth, and technological improvements over a period of, for example, one century. Emission scenarios in which mitigation measures are adopted are referred to as mitigation scenarios.

[*2] Ancillary benefits
Other positive effects produced concurrently with the results of measures to reduce greenhouse gases. For example, the reduction of CO_2 both enhances energy efficiency and mitigates local air pollution, while afforestation to absorb CO_2 also indirectly prevents erosion.

[*3] Annex B countries
Countries with an established reduction target in the Kyoto Protocol. They comprise the developed Western countries, the countries of the former Soviet Union and Eastern Europe (countries in economic transition), Turkey, and Belarus. They are referred to in this way because they are specified in Annex B of the Protocol.

[*4] International emissions trading
A system of trading greenhouse gas emission quotas between countries, centering around CO_2 emissions. A country that has exceeded its target reduction of greenhouse gases can sell the excess share to a country that has not achieved its target, and the purchasing country will be considered to have reduced emissions by the amount of the share purchased. Trading among Annex B countries is approved under the Kyoto Protocol.

[*5] Marginal cost
The additional reduction cost resulting from one unit of increase in the reduction amount. For example, when an overall reduction of 10 tons of CO_2 is necessary, and an additional cost of 10,000 yen is required when the reduction amount is increased from 9 tons to 10 tons, the marginal cost at the time of the 10-ton reduction is 10,000 yen/tC.

[*6] Carbon leakage
Even though CO_2 emissions are reduced in the developed countries, emissions in the developing countries may increase due to foreign trade and capital transfers. This term is

used since it appears as if CO_2 is "leaking" from the developed countries to the developing countries.

[*7] Kyoto Mechanisms

International mechanisms approved to promote efficient achievement of the greenhouse gas emission targets prescribed in the Kyoto Protocol while minimizing reduction costs, by enabling countries' reduction efforts to complement each other. There are three components: international emissions trading, joint implementation, and the Clean Development Mechanism (CDM).

[*8] Clean Development Mechanism (CDM)

A system under which developed countries can invest in greenhouse gas emission reduction projects in developing countries, and apply part of those emission reductions to their own amount of reduction.

[*9] Hybrid car

A car equipped with both an internal combustion engine and an electric motor. The electric motor is mainly used when accelerating, and the engine mainly when driving at a steady speed. The battery is charged by the energy released during deceleration. This system greatly improves fuel consumption by using the characteristics of both propulsion methods in a complementary manner.

[*10] Fuel cell technology

A power generation technology in which electricity is produced by a chemical reaction of hydrogen and oxygen. Since it is a decentralized power-generation system, there is no loss of power transmission, and the heat generated at the time of power generation can be used, thus enhancing energy efficiency. Water is the only substance discharged as a result of energy use, so the burden on the environment is very small.

[*11] CO_2 underground storage technology

Technology to fix CO_2 underground using abandoned mines and oil wells. Technologies to absorb and reduce CO_2 by injecting it deep underground are currently under study.

[*12] Biomass

Plant resources such as wood, grains, and synthetic oils processed from them. Although CO_2 is emitted by the combustion of trees and grains, atmospheric CO_2 is fixed by photosynthesis when they grow. Therefore, there is no overall increase in atmospheric CO_2.

[*13] UNEP

The United Nations Environment Programme. This is an organization of the UN established by a resolution of the General Assembly in 1972. Based on the Declaration on the Human Environment, it supervises various activities of UN organizations related to the environment. It is headquartered in Nairobi.

[*14] COP

Conference of the Parties to the United Nations Framework Convention on Climate Change. The framework and rules for preventing global warming are discussed at this conference. The Kyoto Protocol was adopted at the third such conference, COP3.

[*15] ppmv
Parts per million by volume, a unit of concentration.

[*16] IPCC Tokyo Workshop
A workshop held in Tokyo on the utilization of integrated assessment models, which assess a series of processes in an integrated manner, from the emission of greenhouse gases to climate change and impact. This gathering provided the first opportunity for full-scale debates with researchers from developing countries.

[*17] Objective defined in Article 2 of the Framework Convention on Climate Change
The ultimate objective of the Convention is "stabilization of greenhouse gas concentrations in the atmosphere at a level that would prevent dangerous anthropogenic interference with the climate system."

REFERENCES

Part 1

"Basic Plan for Science and Technology" adopted by the Cabinet Meeting in March 2001.
"Promotional Strategies in Prioritized Fields based on the Basic Plan for Science and Technology" published by the Council for Science and Technology Policy in September 2001.

Part 2

Chapter 1

Feely, R.A., C.L. Sabine, T. Takahashi, and R. Wanninkhof, 2001: Uptake and storage of carbon dioxide in the oceans: The global CO_2 survey. Oceanography, 14, 18-32.
IPCC, 1996: Climate Change 1995: The Science of Climate Change. J.T. Houghton et al. ed., Cambridge Univ. Press, Cambridge, 572pp.
IPCC, 2001: Climate Change 2001: The Scientific Basis. J.T. Houghton et al. ed., Cambridge University Press, Cambridge, 881pp.
Japan Meteorological Agency, 2002: Climate of Japan in 20th century. 116pp. (in Japanese)
Japan Meteorological Agency, 2003: Annual Report on Atmospheric and Marine Environment Monitoring No.3 Observation Results for 2003. 157pp. (in Japanese)
Joyce, T.M. and P. Robbins, 1996: The long-term hydrographic record at Bermuda. Journal of Climate, 9, 3121-3131.
Karl, D., J.E. Dore, R. Likas, A.F. Michaels, N.R. Bates, and A. Knap, 2001: The long-term picture: The U.S. JGOFS time-series programs. Oceanography, 14, 6-17
Kawamura, K., T. Nakazawa, S. Aoki, S. Sugawara, Y. Fujii, and O. Watanabe, 2003: Atmospheric CO_2 variations over the last three glacial-interglacial climatic cycles deduced from the Dome Fuji deep ice core, Antarctica using a wet extraction technique. Tellus, 55B, 126-137.
Machida, T., 1990: A study of concentration variations of greenhouse gases in ancient air using ice core analysis. Doctoral thesis of Tohoku University, 147pp.
Maksyutov, S., T. Machida, H. Mukai, P.K. Patra, T. Nakazawa, G. Inoue, and Transcom-3

Modelers, 2003: Effect of recent observations on Asia CO_2 flux estimates by transport model inversions. Tellus, 55B, 522-529.

Mikami, T. and N. Ishiguro, 1998: Past 550 years climate change reconstructed using ice formation of Lake Suwa. Meteorological Research Note, 191, 73-84. (in Japanese)

Ministry of the Environment, 2001: Annual Report on Observation Results of Ozone Layer for 2000. 45pp. (in Japanese)

Oikawa, T and A. Ito, 2001: Modeling Carbon Dynamics of Terrestrial Ecosystems in Monsoon Asia. In Present and Future of Modeling Global Environmental Change: Toward Integrated Modeling. T. Matsuno and H. Kida eds., Terra Scientific Publishing Company, Tokyo, 207-219.

Rayner, P., I. Enting, R. Francy, and R. Langenfelds, 1999: Reconstructing the recent carbon cycle from atmospheric CO_2, $\delta^{13}C$ and O_2/N_2 observations. Tellus, 51B, 213-232.

Ronald, H., 1998: FLUXNET and the Framework Convention on Climate Change (FCCC). BAHC NEWS, No.6, 11-13.

Tans, P.P., P.S. Brkwin, L. Bruhwiler, T.J. Conway, E.J. Dlugokencky, D.W. Guenther, D.R. Kitzis, P.M. Lang, K.A. Masarie, J.B. Miller, P.C. Novelli, K.W. Thoning, B.H. Vaughn, J.W.C. White, and C. Zhao, 2001: Carbon Cycle. Climate Monitoring and Diagnostic Laboratory Summary Report No.26, 28-50.

Tohjima, Y., 2000: Method for measuring changes in the atmospheric O_2/N_2 ratio by gaschromatograph equipped with a thermal conductivity detector. Journal of Geophysical Research, 105, D11, 14575-14584.

Yamamoto, S., S. Murayama, N. Saigusa, and H. Kondo, 1999: Seasonal and inter-annual variation of CO_2 flux between a temperate forest and the atmosphere in Japan. Tellus, 51B, 402-413.

Zeng, J., Y. Nojiri, Y. Fujinuma, P. Murphy, and C.S. Wong, 2002: A comparison of delta PCO_2 distributions in the northern North Pacific using results from a commercial vessel in 1995-1999. Deep-Sea Research II, 49, 5303-5315.

Chapter 2

Abe, A., 1997: Study of climate change such as global warming by using coupled atmosphere-ocean general circulation models. In: Frontiers of climate studies, Climate System Study, Vol.2, Chapter 9, 117-130, Center for Climate System Research, Tokyo. (in Japanese)

Emori, S., T. Nozawa, A. Abe-Ouchi, A. Numaguti, M. Kimoto and T. Nakajima, 1999: Coupled ocean-atmosphere model experiments of future climate change with an explicit representation of sulfate aerosol scattering. J. Met. Soc. Japan, 77, 1299-1307.

Emori, S., T. Nozawa, A. Numaguchi and I. Uno, 2000: A regional climate change projection over East Asia. Preprint Volume of the 11th Symposium on Global Change Studies, 9-14 January 2000, Long Beach, Cal., 15-18.

IPCC, 1990: Climate Change, The IPCC Scientific Assessment. J.T. Houghton et al. ed., Cambridge Univ. Press, Cambridge, 365pp.

IPCC, 1996: Climate Change 1995: The Science of Climate Change. J.T. Houghton et al. ed., Cambridge Univ. Press, Cambridge, 572pp.

IPCC, 2001a: Climate Change 2001: The Scientific Basis. J.T. Houghton, et al. ed., Cambridge University Press, Cambridge, 881pp.

IPCC, 2001b: Climate Change 2001: Impacts, Adaptation, and Vulnerability. J.J. McCarthy, et al. ed., Cambridge University Press, Cambridge, 1032pp.

Japan Meteorological Agency, 1996: Information of Global Warming, Vol.1. National Printing Bureau, Tokyo. (in Japanese)

Japan Meteorological Agency, 1998: Information of Global Warming, Vol.2. National Printing Bureau, Tokyo. (in Japanese)

Japan Meteorological Agency, 1999: Information of Global Warming, Vol.3. National Printing Bureau, Tokyo. (in Japanese)

Japan Meteorological Agency, 2001: Information of Global Warming, Vol.4. National Printing Bureau, Tokyo. (in Japanese)

Kato, H., K. Nishizawa, H. Hirakuchi, S. Kadokura, N. Oshima and F. Giorgi, 2001: Performance of RegCM2.5/NCAR-CSM nested system for the simulation of climate change in East Asia caused by global warming. J. Meteor. Soc. Japan, 1, 99-121.

Leggett, J., W.J. Pepper, and R.J. Swart, 1992: Emissions Scenarios for IPCC: an update. In: Climate Change 1992. The Supplementary Report to the IPCC Scientific Assessment [Houghton, J.T., B.A. Callander, and S.K. Varney (eds.)]. Cambridge University Press, Cambridge, United Kingdom and New York, NY, USA, pp. 69-95.

Lomborg, B., 2001:The Skeptical Environmentalist, Cambridge University Press, London, 540pp.

Manabe, S., and R.T. Wetherald, 1967: Thermal equilibrium of atmosphere with a given distribution of relative humidity, J. Atmos. Sci., 24, 241-259.

Nakicenovic, N., J. Alcamo, G. Davis, B. de Vries, J. Fenhann, S. Gaffin, K. Gregory, A. Gru"bler, T.Y. Jung, T. Kram, E.L. La Rovere, L. Michaelis, S. Mori, T. Morita, W. Pepper, H. Pitcher, L. Price, K. Raihi, A. Roehrl, H.-H. Rogner, A. Sankovski, M. Schlesinger, P. Shukla, S. Smith, R. Swart, S. van Rooijen, N. Victor, and Z. Dadi, 2000: Emissions Scenarios. A Special Report of Working Group III of the Intergovernmental Panel on Climate Change. Cambridge University Press, Cambridge, United Kingdom and New York, NY, USA, 599 pp.

Noda, A., 2000: Projection of global climate change due to global warming, Tenki, 47, 702-708. (in Japanese)

Nozawa, T., S. Emori, A. Numaguti, Y. Tsushima, T. Takemura, T. Nakajima, A. Abe-Ouchi and M. Kimoto, 2001: Projections of future climate change in the 21st century simulated by the CCSR/NIES CGCM under the IPCC SRES scenarios, In: Present and Future of Modelling Global Environmental Change Toward Integrated Modelling, T. Matsuno and H. Kida (ed), 15-28, Terra Scientific Publishing Company, Tokyo.

Sato, Y., 2000: Projection of regional climate change over Japan due to global warming. Tenki, 47, 708-716. (in Japanese)

Sumi, A., 2001: Global environmental issues in the 21st century as viewed from climate system science. Environment and Information Science, Vol. 30, No. 1, 18-21. (in Japanese)

Tsushima, Y., and S. Manabe, 2001: Influence of cloud feedback on annual variation of global mean surface temperature. Journal of Geophysical Research (Atmosphere), 106-D19, 22,635-22,646.

Chapter 3

Alward, R.D. and J.K. Detling, 1999: Grassland vegetation changes and nocturnal global warming. Science, 283, 229-231.

Arendt, A.A., K.A. Echelmeyer, W.D. Harrison, C.S. Lingle, and V.B. Valentine, 2002: Rapid wastage of Alaska glaciers and their contribution to rising sea level. Science, 297, 382-386.

Bradley, N.L., A.C. Leopold, J. Ross, and W. Huffaker, 1999: Phenological changes reflect climate change in Wisconsin. Proceedings of the National Academy of Sciences of the United States of America, 96, 9701-9704.

Brown, J.H., T.J. Valone, and C.G. Curtin, 1997: Reorganization of an arid ecosystem in response to recent climate change. Proceedings of the National Academy of Sciences of the United States of America, 94, 9729-9733.

Dettinger, M.D. and D.R. Cayan, 1995: Large-scale atmospheric forcing of recent trends toward early snowmelt runoff in California. Journal of Climate, 8, 606-623.

Fitter, A.H. and R.S.R. Fitter, 2002: Rapid changes in flowering time in British plants. Science, 296, 1689-1691.

Gleick, P., 2000: The World Water 2000-2001. Island Press, 315pp.

Gonzalez, P., 2001: Desertification and a shift of forest species in the West African Sahel. Climate Research, 17(2), 217-228.

Grabherr, G., M. Gottfried, and H. Pauli, 1994: Climate effects on mountain plants. Nature, 369, 448.

Hamburg, S.P. and C.V. Cogbill, 1988: Historical decline of red spruce population and climatic warming. Nature, 331, 428-431.

Hasenauer, H., R.R. Nemani, K. Schadauer, and S.W. Running, 1999: Forest growth response to changing climate between 1961 and 1990 in Austria. Forest Ecology and Management, 122, 209-219.

Harasawa, H. and S. Nishioka, 2003. Global Warming Impacts on Japan, kokonshoin, 411pp. (in Japanese)

Horie,T., H.Nakagawa, M.Ohnishi and J.Nakano, 1995 : Rice production in Japan under current and future climates. In:Mathews. R.B. et al. (eds.) Modelling the Impacts of Climate Change on Rice in Asia, CAB International, Oxon, U.K. 143-164.

IPCC, 1994: IPCC Technical Guidelines for Assessing Climate Change Impacts and Adaptation. T.R. Carter et al. ed., University College London and Center for Global Environmental Research, 59pp.

IPCC, 1996: Climate Change 1995 - Impacts, Adaptations, and Mitigation of Climate Change: Scientific-Technical Analyses. R. T. Watson et al. ed., Cambridge Univ. Press, Cambridge, 878pp.

IPCC, 1998 : The Regional Impacts of Climate change -An Assessment of Vulnerability. Cambridge Univ. Press, Cambridge, 517pp.

IPCC, 2001: Climate Chang 2001 - Impacts, Adaptation, and Vulnerability. J. J. McCarthy et at. ed., Cambridge Univ. Press, 1032pp.

Ishigami, Y., Y. Shimizu, and K. Omasa, 2001 : Prediction of changes in vegetation distribution in Japan using process based model and GCM data. Proceeding of LUCC Symposium 2001.

Kainuma, M., Y. Matsuoka, and T. Morita, 2002: Climate Policy Assessment Asia-Pacific Integrated Modeling. Spriger, 402pp.

Masuda, K., M. Yoshino, and K. Park, 1999: Detection of global warming impacts using biophenology, Global Environment, 4(1&2), 91-103. (in Japanese)

Menzel, A. and P. Fabian, 1999: Growing season extended in Europe. Nature, 397, 659.

Mimura, N. and E. Kawaguchi, 1996 :Responses of coastal topography to sea-level rise. Proc. of 25th ICCE, 1161-1165.

Ministry of the Environment, 2001: Climate Change 2001 IPCC Third Assessment Report – Summary for Policy Makers, 91pp. (in Japanese)

NASA, 2002: http://www.gsfc.nasa.gov/

NAST (National Assessment Synthesis Team), 2001: Climate Change Impacts on the United States. The Potential Consequences of Climate Variability and Change. 612pp.

National Land Preservation Study Group for Sea-Level Rise Accompanying Global Warming, 2002: Report of the Study Group, Ministry of Land, Infrastructure and Transport, 12pp., with Appendix 68pp. (in Japanese)

Ohosaka, S., 1996: Climate and electricity operation – meteorological observation and systematic operation in electric power supply, Electrical Review, 12, 15-18. (in Japanese)

Parmesan, C., 1996: Climate and species range. Nature, 382, 765-766.

Peñueleas, J. and I. Filella, 2001: Response to a warming world. Science, 294, 793-795.

Research Committee on National Land Conservation to Sea Level Rise due to Global Warming, 2002: Report of Research Committee on National Land Conservation to Sea Level Rise due to Global Warming. (in Japanese)

UNEP, 2002: http://climatechange.unep.net/

USCSP, 1999: Climate Change Mitigation, Vulnerability, and Adaptation in Developing and Transition Countries. 96p+Appendix.

Chapter 4

Amano, H., 2002: Socio-economic dimension of global environmental problems, In Morita, T. and H. Amano, eds. Global Environmental Problems and Global Community, Environmental and Economic Policy, Iwanami Publishing CO., Vol.6, 9-36. (in Japanese)

Alcamo, J., A. Bouwman, J. Edmonds, A. Grubler, T. Morita, and A. Sugandhy, 1995: An Evaluation of the IPCC IS92 Emission Scenarios. In Climate Change 1994, Radiative Forcing of Climate Change and an Evaluation of the IPCC IS92 Emission Scenarios. Cambridge University Press, Cambridge.

Aoyama, T., 1997: Energy Perspectives in Asia, The Future of Nuclear Power Generation. Energy in Japan, No.145, 42-50.

Fujii, Y., and K. Yamaji, 1998: Assessment of Technological Options in the Global Energy System for Limiting the Atmospheric CO_2 Concentration. Environmental Economics and Policy Studies, 1(2), 113-139.

Fujime, K, 1998: Japan s Latest Long-term Energy Supply and Demand Outlook From Kyoto Protocol s Perspectives. Energy in Japan, No. 153, 1-9

Global Environment Subcommittee, Central Environment Council, 2001: Intermediate Report on Subcommittee on Scenarios to Attain the Target. (in Japanese) http://www.env.go.jp/council/06earth/r062-01/index.html

Institute for Global Environmental Strategies, 2000: National climate policy in Japan, IGES Open Forum on Global Warming Policy, 18pp. (in Japanese)

Institute for Global Environmental Strategies, 2002: Response policy to climate change issues – Proposal on national policy portfolio. (in Japanese)

IPCC, 2001: Climate Change 2001: Mitigation. B. Metz et al. ed., Cambridge University Press, Cambridge, 752pp. (http://www.grida.no/climate/ipcc_tar/wg3/index.htm)

Ishitani, H., Y. Baba and O. Kobayashi, 2000a: Well to wheel energy efficiency of fuel cell electric vehicles by various fuels. Japan Society of Energy and Resources, Energy and Resources, 21(5), 417-425.

Ishitani, H., Y. Baba, and R. Matsuhashi, 2000b: Evaluation of Energy Efficiency of a Commercial HEV, Prius at City Driving. Department of Geosystems Engineering,

University of Tokyo, Japan.

Iwahashi, T., 1998: Basic Design of a Domestic System of Greenhouse Gas Emissions Trading. Osaka Law Review, 48(3), 857-890.

Japan Energy Conservation Center, 1997: Energy Conservation Handbook, Japan.

Japan Resources Association, 1994: Life cycle energy in domestic life. Anhorume Co Ltd., Japan.

Jiang, K., T. Morita, T. Masui, and Y. Matsuoka, 2000: Global Long-term GHG Mitigation Emission Scenarios based on AIM. Environmental Economics and Policy Studies, 3(2), 239-254.

Jiang, K., X. Hu, Y. Matsuoka, and T. Morita, 1998: Energy Technology Changes and CO_2 Emission Scenarios in China. Environmental Economics and Policy Studies, 1(2), 141-160.

Kainuma, M., Y. Matsuoka, and T. Morita, 1999a: Analysis of Post-Kyoto Scenarios: The Asian Pacific Integrated Model. Energy Journal, Kyoto Special Issue, 207-220.

Kainuma, M., Y. Matsuoka, T. Morita, and G. Hibino, 1999b: Development of an End-Use Model for Analyzing Policy Options to Reduce Greenhouse Gas Emissions. IEEE Transactions on Systems, Man, and Cybernetics -Part C: Applications and Reviews, 29(3 -August), 317-323.

Kainuma, M., Y. Matsuoka, and T. Morita, 2000: The AIM/End-Use Model and its Application to Forecast Japanese Carbon Dioxide Emissions. European Journal of Operational Research, 392-404.

Kanie, K. 2001: Global environmental diplomacy and national domestic policy – Kyoto Protocol and Netherland's diplomacy and policy, 336pp, Keio University Press. (in Japanese)

Kashiwagi, T., B.B. Saha, D. Bonilla, and A. Akisawa, 1999: Energy Efficiency and Structural Change for Sustainable Development and CO_2 Mitigation. Contribution to the IPCC Expert Meeting "Costing Methodologies", Dept. of Mechanical Systems Engineering, Tokyo University of Agriculture and Technology, Tokyo, June.

Kawashima, Y., 1997: Comparative Analysis of Decision-Making Process of the Developed Countries Towards CO_2 Emission Reduction Target. International Environmental Affairs, 9, 95-126.

Kawashima, Y., 2000: Japan s Decision-Making About Climate Change Problems: Comparative Study of Decisions in 1990 and 1997. Environmental Economics and Policy Studies, 3(1), 29-57.

Kurosawa, A., H. Yagita, Z. Weisheng, K. Tokimatsu, and Y. Yanagisawa, 1999: Analysis of Carbon Emission Stabilization Targets and Adaptation by Integrated Assessment Models. Energy Journal, Kyoto Special Issue, 157-175.

Matsuo, N., 1998: Points and proposals for the emission trading regime of climate change -

For designing future systems (Version 2). IGES, Shonan Village, Japan.

Matsuo, N., 1999: Baselines as the critical issue of CDM -possible pathways to standardisation, In Proceedings CDM workshop - workshop on baseline for CDM, NEDO/GISPRI, (eds.), Tokyo, 9-22.

Matsuoka, Y., 2000: Development of a Stabilization Scenario Generator for Long-term Climatic Assessment. Environmental Economics and Policy Studies, 3 (2), 255-266.

Matsuoka, Y., M. Kainuma, and T. Morita, 1995: Scenario Analysis of Global Warming Using the Asian Pacific Integrated Model (AIM). Energy Policy, 23(4-5), 357-371.

Miyamoto, K. (ed), 1997: Renewable biological systems for alternative sustainable energy production. FAO Agricultural Series Bulletin 128.

Mori, S. 2000: Effects of Carbon Emission Mitigation Options under Carbon Concentration Stabilization Scenarios. Environmental Economics and Policy Studies, 3(2), 125-142.

Mori, S., and M. Takahashi, 1998: An Integrated Assessment Model for New Energy Technologies and Food Production -An Extension of the MARIA Model. International Journal of Global Energy Issues, 11(1-4), 1-17.

Morita, T., and H. Lee, 1998a: IPCC SRES database, Version 1.0, Emissions Scenario Database prepared for IPCC Special Report on Emission Scenarios (online: http://www-cger.nies.go.jp/cger-e/db/ipcc.html)

Morita, T., and H. Lee, 1998b: IPCC Emission Scenarios Database. Mitigation and Adaptation Strategies for Global Change, 3(2-4), 121-131.

Morita, T., N. Nakicenovic and J. Robinson, 2000a: Overview of Mitigation Scenarios for Global Climate Stabilization based on New IPCC Emission Scenarios (SRES). Environmental Economics and Policy Studies, 3(2), 65-88.

Morita, T., N. Nakicenovic, and J. Robinson, 2000b: The Relationship between Technological Development Paths and the Stabilization of Atmospheric Greenhouse Gas Concentrations in Global Emissions Scenarios. CGER Report (CGER-I044-2000), Center for Global Environmental Research, National Institute for Environmental Studies.

Nakicenovic, N., J. Alcamo, G. Davis, H.J.M. de Vries, J. Fenhann, S. Gaffin, K. Gregory, A. Grubler, T.Y. Jung, T. Kram, E.L. La Rovere, L. Michaelis, S. Mori, T. Morita, W. Papper, H. Pitcher, L. Price, K. Riahi, A. Roehrl, H-H. Rogner, A. Sankovski, M. Schlesinger, P. Shukla, S. Smith, R. Swart, S. van Rooijen, N. Victor, and Z. Dadi, 2000: Special Report on Emissions Scenarios. Intergovernmental Panel on Climate Change, Cambridge University Press, Cambridge, 599.

NEDO, 1999: Fluidised bed combustion. Technical Brief from Residua & Warmer Bulletin [http://www.residua.com/wrftbfbc.html]

Ogawa, J., B. Bawks, N. Matsuo, and K. Ito, 1998: The Characteristics of Transport Energy Demand in the APEC Region. Asia Pacific Energy Research Center, Tokyo.

Ozaki, M., 1997: CO_2 Injection and Dispersion in Mid-Ocean Depth by Moving Ship. Waste

Management, 17(5/6), 369-373.

Rana, A. and T. Morita, 2000: Scenarios for Greenhouse Gas Emissions Mitigation: A Review of Modeling of Strategies and Policies in Integrated Assessment Models. Environmental Economics and Policy Studies, 3(2), 267-289.

Sato, O., M. Shimoda, K. Tatemasu, and Y. Tadokoro, 1999: Roles of Nuclear Energy in Japan s Future Energy System. English summary of the JAERI technical paper 99-015, March.

Sato, T., 1999: Can we make the Biological Impacts of CO_2 Negligible by Dilution? In Proceedings of the 2nd International Symposium on Ocean Sequestration of Carbon Dioxide. NEDO, 21-22 June 1999, Tokyo, Japan.

Yagi, K.H., H. Tsuruta, and K. Minami, 1997: Possible options for mitigating methane emission from rice cultivation. Nutrient Cycling in Agroeco, 49, 213-220.

Yamaji, K., 1998: A Study of the Role of End-of-pipe Technologies in Reducing CO_2 Emissions. Waste Management, 17(5/6), 295-302.

Yamaji, K., J. Fujino, and K. Osada, 2000: Global Energy System to Keep the Atmospheric CO_2 Concentration at 550 ppm. Environmental Economics and Policy Studies, 3(2), 159-172.

Yamasaki, E. and T. Tominaga, 1997: Evolution of an ageing society and effect on residential energy demand. Energy Policy, 25 (11), 903-912.

ABBREVIATIONS

A-GCM: Atmospheric General Circulation Model
AO-GCM: Atmosphere-Ocean General Circulation Model
AIACC: Assessment of Impacts and Adaptations to Climate Change
AsiaFlux: Asia Flux Network
BAHC: Biospheric Aspects of the Hydrological Cycle
BATS: Bermuda Atlantic Time Series
CANZ: Canada, Australia, New Zealand
CCSM: Community Climate System Model
CCSR/NIES: Center for Climate System and Research / National Institute for Environmental Studies
CDM: Clean Development Mechanism
CFC: Chlorofluorocarbon
CGCM: Coupled General Circulation Model
CGER: Center for Global Environmental Research (NIES, Japan)
CLIVAR: Climate Variability and Predictability (WCRP)
CliC: Climate and Cryosphere (WCRP)
COADS: Comprehensive Ocean-Atmosphere Data Set
COD: Chemical Oxygen Demand
COP: the Conference of the Parties of the United Nations Framework Convention on Climate Change
CMIP: Coupled Model Intercomparison Project (WGCM)
CSTP: Council for Science and Technology Policy (Cabinet Office, Japan)
ENSO: El Niño / Southern Oscillation
ERBE: Earth Radiation Budget Experiment
FRSGC: Frontier Research System for Global Change (Japan)
GCM: General Circulation Model
GCTE: Global Change and Terrestrial Ecosystems
GDP: Gross Domestic Product
GEWEX: Global Energy and Water Cycle Experiment (WCRP)
GFDL: Geophysical Fluid Dynamics Laboratory (US)
GISS: Goddard Institute for Space Studies (US)
GPP: Gross Primary Production
HCFC: Hydrochlorofluorocarbon
HOT: Hawaii Ocean Time Series

ICSU: International Council for Science
IGAC: International Global Atmospheric Chemistry Project
IGBP: International Geosphere-Biosphere Programme
IGY: International Geophysical Year
IPCC: Inter-governmental Panel on Climate Change (UNEP-WMO)
JAXA: Japan Aerospace Exploration Agency (since October 2003)
JGOFS: Joint Global Ocean Flux Study
KNOT: Kyodo North Pacific Ocean Time Series (JGOFS)
MRI: Meteorological Research Institute (Japan)
NASA: National Aeronautics and Space Administration (US)
NASDA: National Space Development Agency (Japan) (JAXA, since October 2003)
NAST: National Assessment Synthesis Team (US)
NCAR: National Center for Atmospheric Research (US)
NCEP: National Centers for Environmental Prediction (US NOAA)
NEP: Net Ecosystem Production
NIES: National Institute for Environmental Studies (Japan)
NGO: Non Governmental Organization
NOAA: National Oceanic and Atmospheric Administration (US)
NPP: Net Primary Production
OCTS: Ocean Color Temperature Scanner
ODA: Official Development Assistance
PRISM: Program for Integrated Earth System Modeling
SeaWiFS: Sea-viewing Wide Field-of-view Sensor
SPARC: Stratospheric Processes And their Role in Climate (WCRP)
SRES: IPCC Special Report on Emission Scenario
START: Global Change System for Analysis, Research and Training (IHDP-IGBP-WCRP-APN-ENRICH-IAI)
TAR: IPCC Third Assessment Report
TWAC: Third World Academy of Science
UKMO: United Kingdom Meteorological Office
UNEP: United Nations Environment Programme
USCSP: United States Country Study Program
WCRP: World Climate Research Program
WDCGG: World Data Center for Greenhouse Gases
WGCM: Working Group on Climate Model
WMO: World Meteorological Organization

AUTHORS
(Parts of the book written by the author are identified in parentheses)

Atsunobu Ichikawa (Foreword)
Born in 1930.
Special Advisor to the President, Japan Science and Technology Agency; Professor Emeritus, Tokyo Institute of Technology.
Field of Interest: System science and environmental science

Shiro Ishii (Preface)
Born in 1935.
Professor Emeritus of the University of Tokyo; Former member of the Council for Science and Technology Policy (January 6, 2001 -January 5, 2003).
Field of Interest: Japanese regal history

Kenji Ohkuma (Preface)
Born in 1946.
Director General, Bureau of Science and Technology Policy, Cabinet Office.

Yasuhiro Sasano (Part I)
Born in 1952.
Director for Environment and Energy, Bureau of Science and Technology Policy, Cabinet Office.
Field of Interest: Atmospheric physics and environmental science

Makoto M. Watanabe (Part I)
Born in 1948.
Director of Environmental Biology Division, National Institute for Environmental Studies; Former Director for Environment and Energy, Bureau of Science and Technology Policy, Cabinet Office (January 6, 2001 - July 1, 2002).
Field of Interest: Biodiversity science

Isao Koike (1-1, 1-2, 1-4-3, 1-4-4, Ch.1, Part II)
Born in 1944.
Director and Professor, Ocean Research Institute, The University of Tokyo.
Field of Interest: Marine biogeochemistry

Gen Inoue (1-3, Ch.1, Part II)
Born in 1945.
Director, Center for Global Environmental Research; Assistant Leader, Climate Change Research Project, National Institute for Environmental Studies.
Field of Interest: Atmospheric chemistry and monitoring

Susumu Yamamoto (1-4-1, 1-4-2, Ch.1, Part II)
Born in 1945.
Deputy Director, Institute for Environmental Management Technology, National Institute of Advanced Industrial Science and Technology.
Field of Interest: Atmospheric environmental science and field-based measurements of the carbon cycle

Akira Noda (2-1, 2-2, 2-3-1, 2-4-2, 2-4-3, Ch.2, Part II)
Born in 1949.
Head, The Fourth Laboratory, Climate Research Department, Meteorological Research Institute.
Field of Interest: Global warming projections based on global climate models

Taroh Matsuno (2-1, 2-4-1, 2-4-2, 2-4-4, Ch.2, Part II)
Born in 1934.
Director General, The Frontier Research System for Global Change; Professor Emeritus, The University of Tokyo
Field of Interest: Meteorology and climate dynamics

Akimasa Sumi (2-3-2, 2-4-2, Ch.2, Part II)
Born in 1948.
Director and Professor, Center for Climate System Research, The University of Tokyo
Field of Interest: Meteorology

Hiroki Kondo (Ch.2, Part II)
Born in 1941
Research Supervisor, Integrated Modeling Research Program, Frontier Research System for Global Change
Field of Interest: Projection of global warming with its application to regional scales

Hideo Harasawa (3-1, Ch.3, Part II)
Born in 1954.

Head, Environmental Planning Section, Social and Environmental Systems Division, National Institute for Environmental Studies

Field of Interest: Environmental engineering and impact assessment of global warming

Nobuo Mimura (3-2, Ch.3, Part II)
Born in 1949.

Professor, Center for Water Environment Studies, Ibaraki University.

Field of Interest: Global environmental engineering and coastal engineering

Yousay Hayashi (3-2, Ch.3, Part II)
Born in 1948.

Director, Department of Global Resources, National Institute for Agro-Environmental Sciences.

Field of Interest: Agricultural climatology

Tsuneyuki Morita (Ch.4, Part II)
Born in 1950; Died in 2003

Director, Social and Environmental Systems Division, National Institute for Environmental Studies.

Field of Interest: Environmental economics, Policy science, Global modeling

Shuzo Nishioka (Ch.4, Part II)
Born in 1939.

Executive Director, National Institute for Environmental Studies

Field of Interest: Environmental systems analysis, Environmental policy

Rie Watanabe (Ch.4, Part II)
Born in 1968.

Research Associate, Institute for Global Environmental Strategies; Visiting Research Fellow, Environmental Law Center, IUCN.

Field of Interest: Environmental law and policy

Tribute to Dr. Tsuneyuki Morita

Dr. Tsuneyuki Morita passed away on 4 September, 2003, at the age of 53.

In 1988 he commenced his research career at the National Institute for Environmental Studies. This was followed by remarkable achievements in a variety of research and associated activities in environmental economics and related policy fields. Among them, the most noteworthy achievement is his contribution to global warming policies, through IPCC, academic societies and governmental committees. He demonstrated outstanding leadership as Director of the Social and Environmental Systems Division, National Institute for Environmental Studies. Since 2002 he also served as Coordinator of the Global Warming Research Initiative, organized by the Council for Science and Technology Policy, Cabinet Office. He was therefore an author of the Japanese version of this book "Global Warming-The Research Challenges".

It is with deep sorrow that we lost the most outstanding advocate and leader for the promotion of the global warming policy research and, more personally, a fellow author of this book and a good friend. All we can do now is acknowledge with respect and gratitude his significant contributions to climate change science and policy, and pray that he rests in peace.

INDEX

actor-based approach, 130
adaptation, 89
administrative reform, 5
aerosols, 19, 33, 60, 72
afforestation, 127
Agenda 21, 3
agriculture, 93
aircraft data, 44
air-sea P_{CO_2}, 49
air-sea CO_2 flux, 48
amount of chlorophyll, 51
ancillary benefits, 116
Annex B countries, 122
anomaly of land surface temperature, 30
anomaly of sea-surface temperature, 30
area of sea ice, 104
Arrhenius, 56
AsiaFlux, 44

BATS, 53
bicarbonate ions, 46
biodiversity, 104
biological pump, 46
boundary conditions, 77
brackish lakes, 104
bottom-up model, 124
buoy system for monitoring El Niño, 53
buoy systems, 53

capacity for absorption and repair, 27
carbon accumulation, 37
carbon cycle models, 34
carbon cycle, 37
carbon fixation technologies, 126
carbon leakage, 123
CCSM, 65
Central Research Institute of Electric Power Industry (CRIEPI), 78
CH_4, 20
changes in oxygen concentration, 34
Chemical Substance Risk Management, 8
Clean Development Mechanism (CDM), 123
CliC, 64
climate change and future prediction, 10
climate change, 4
climate model, 59
climate sensitivity experiments, 68
climate system, 58
CLIVAR, 64
cloud radiative forcing, 76
CO_2 absorption, 40
CO_2 equivalent, 3
CO_2 fertilization effect, 37
CO_2 flux, 35, 39, 40
CO_2 increase rate in the tropics, 29
CO_2 underground storage technology, 124
CO_2, 20

coastal disaster prevention, 110
Comprehensive Ocean-Atmosphere
 Data Set (COADS), 24
computer models, 117
consumption patterns, 111
convective precipitation, 69
COP3, 27
coral reefs, 95, 105
cost of long-term mitigation
 measures, 121
Council for Science and Technology
 Policy (CSTP), 3, 5, 14
countries in economic transition,
 135
coupled atmosphere-ocean model,
 69
cumulus convection, 83

demand for power, 111
dengue, 95
deteriorated water quality, 104
development scenarios, 118
direct effect, 72
Dome Fuji in Antarctica, 27
downscaling, 77
drought, 107

Earth Simulator, 67, 80
earth s surface temperature, 20
Eco-Harmonious River Basin and
 Urban Area Regeneration, 8
economic assessments, 133
El Niño and Southern Oscillation
 (ENSO), 69
El Niño, 23, 62
emission scenarios, 117

emission suppression technology, 10
enhanced oil recovery, 126
environmental democracy, 130
equilibrium experiments, 63
eutrophication, 95
evacuation corridors, 107
examples of detected global
 warming impacts, 87
extreme climate events, 90

feedback effects, 74
fixing / sequestrating technology, 10
flooding, 107
flowering date of the Japanese
 cherry (*Prunus yedoensis*), 101
food security, 94, 108
forest ecosystem, 37
forest sequestration models, 42
forests, 102
fossil fuel consumption, 34
Fourier, 55
Frontier Research System for Global
 Change (FRSGC), 74
fuel cell technology, 124

gas chromatography, 36
GCTE, 39
GDP loss, 122
GEF, 98
GEWEX, 64
global monitoring, 19
global warming, 19
global warming issue, 55
global warming pattern, 70
Global Warming Research Initiative,
 4

Global Warming Response Policies, 115
Global Water Cycle, 8
GPP, 38
greenhouse effect, 74
greenhouse gases, 19, 33

Halogenated hydrocarbon, 31
harbor facilities, 110
Hateruma Island, 29
Hateruma monitoring station, 36
heat island phenomenon, 96
HOT, 53
human health, 111
hybrid cars, 124

ice sheets, 24
IGBP, 39
impact / risk study, 11
impact risk, 91
impact studies, 98
impacts on Japan, 100
impacts on water resources, 92
indirect effect, 72
infectious diseases, 111
Information on Global Warming, 69
initiative promotion system, 14
initiatives, 6,7,13
inorganic carbon, 46
integrated Earth system model, 82
interests between North and South, 135
intergenerational equity, 134
Intergovernmental Panel on Climate Change (IPCC), 9, 63, 115
international emissions trading, 122

International Geophysical Year, 47
international negotiation processes, 130
inverse model, 44
IPCC Third Assessment Report (TAR), 19, 85
irreversible changes, 88
IS92a, 117

JGOFS, 53
Johannesburg Declaration, 3
joint implementation (JI), 135

KNOT, 53
Kobe Marine Observatory, 24
Kyoto Mechanisms, 123
Kyoto Protocol, 3, 127

La Niña, 23, 62
land surface vegetation, 60
large-scale precipitation, 69
law and economics approach, 134
LCA, 10
life science, 6
long-term emission scenarios, 116
long-term observation and research networks for the amount of CO_2 exchanges (flux), 43

malaria, 95
marginal cost, 122
marine research ship, "Mirai", 53
Mauna Loa, 28

measures to mitigate global
   warming, 113
Meteorological Research Institute
   (MRI), 68, 78
Methane, 31
Minamitorishima, 28
mitigation costs, 116
mitigation scenarios, 116
model intercomparison, 65
moist adiabatic lapse rate, 61
monitoring, 26
monitoring and information
   delivery, 11
monitoring as an indicator of
   change, 99
Monitoring and Process Studies, 12
mountain glaciers, 24
Mt. Pinatubo, 30

nanotechnology, 6
NASA, 52
National Institute for Environmental
   Studies (NIES), 72, 78
natural variability, 69, 70
Net Ecosystem Production (NEP),
   38
net primary productions, 42
Nitrous Oxide, 31
non-governmental organizations
   (NGOs), 3
non-regret measures, 129
NPP, 38, 42
numerical experiment, 63
numerical weather prediction, 66
nutrient, 51
observations using a tower, 39
ocean absorption, 36

ocean color sensor, 51, 52
ocean temperature changes, 22
oceanic carbon cycle, 51
oceanic ice, 24
OCTS, 51
Omiwatari, "the divinity s pathway",
   25
oxygen isotope, 27
oxygen/nitrogen ratio, 36
ozone hole, 31
ozone in the troposphere, 32

parameterization, 59
participation by the developing
   countries, 132
per meg, 36
phenological observations, 101
phytoplankton, 46
planetary boundary layer, 72
policy mix, 128
policy research, 11
policy decision processes, 116, 130
post-SRES scenarios, 121
potential vegetation, 103
precautionary principle, 129
PRISM, 65
Projection Modeling and Climate
   Change Study, 12
Promotional Strategies in Prioritized
   Fields, 6
promotional strategies, 6

Radiative forcing due to changes
   in greenhouse gases from
   pre-industrial time to the present,
   26

Index

radiative forcing, 19, 20
recycling, 127
regional climate models, 77, 78
regional impacts, 97
regular monitoring, 28
Reliability of Carbon-cycle Models, 43
resolution, 65
renewable energy, 126
rice, 108

sand beach erosion, 105
scientific uncertainty, 130
sea ice in the Arctic Sea, 24
SeaWiFS, 51
sea-level rises, 110
second commitment period, 136
Sim-CYCLE, 42
slab ocean model, 69
sources and sinks, 33
Southern Oscillation Index (SOI), 30
SPARC, 64
SRES scenarios, 118
stabilization target, 132
stable concentrations, 99
standard gas, 44
subtropical deep-sea water, 23
sulfuric acid mist, 32
surface temperature in the Northern Hemisphere, 21
Sustainable Co-existence Project on Human, Nature and the Earth, 81
Syowa Station, 28

technological innovation, 133

temperature changes, 22
temperature lapse rate, 61
temperatures of the upper ocean, 23
terrestrial absorption, 36
terrestrial ecosystems, 95
threshold, 113
time-series observations, 53
timing of mitigation measures, 134
transient experiments, 63
troposphere, 58
Tyndall, 55

United Nations Framework Convention on Climate Change (UNFCCC), 3
upwelling in the equatorial zone, 47

value of fishing grounds, 105
vulnerability, 85

Walker Circulation, 62
Waste-Free and Resource Recycling Technologies, 8
water shortages, 93
WCRP, 64
World Summit on Sustainable Development, 3

P_{CO_2}, 46